The Origins of the Horizon in Husserl's Phenomenology

CONTRIBUTIONS TO PHENOMENOLOGY

IN COOPERATION WITH THE CENTER FOR ADVANCED RESEARCH IN PHENOMENOLOGY

Volume 67

Scope

The purpose of the series is to serve as a vehicle for the pursuit of phenomenological research across a broad spectrum, including cross-over developments with other fields of inquiry such as the social sciences and cognitive science. Since its establishment in 1987, *Contributions to Phenomenology* has published nearly 60 titles on diverse themes of phenomenological philosophy. In addition to welcoming monographs and collections of papers in established areas of scholarship, the series encourages original work in phenomenology. The breadth and depth of the Series reflects the rich and varied significance of phenomenological thinking for seminal questions of human inquiry as well as the increasingly international reach of phenomenological research.

For further volumes:
http://www.springer.com/series/5811

Saulius Geniusas

The Origins of the Horizon in Husserl's Phenomenology

Springer

Saulius Geniusas
Philosophy
Chinese University of Hong Kong
Hong Kong, China

ISSN 0923-9545
ISBN 978-94-007-4643-5 ISBN 978-94-007-4644-2 (eBook)
DOI 10.1007/978-94-007-4644-2
Springer Dordrecht Heidelberg New York London

Library of Congress Control Number: 2012943049

Printed on acid-free paper

Springer is part of Springer Science+Business Media (www.springer.com)

Acknowledgments

Philosophical works are not written in solitude. Up to a large degree, they arise out of numerous conversations that the author has had the fortunate to take part in. I am grateful to Nicolas deWarren—one of the most inspiring partners in discussion I have had the fortune of meeting. A deep expression of thanks is due to Claudio Majolino for having read an early draft of this manuscript; conversations with him have had a tremendous impact upon my research. I owe debts of thanks to Mindaugas Briedis, Jagna Brudzinska, Mike Gubser, Sophie Loidolt, Luis Niel, Mark Piper, and Mario Wenning, with whom I have had the fortune to enter into copious philosophical discussions that have always left me enriched, yet always craving for a continuation. I am also grateful to Rashmika Pandya, in whose company my journey in phenomenology began, and to Dominic Lavoie, Nishad Patnaik, and Moris Stern in whose company this journey continued. Lester Embree and Roberto Walton have read early drafts of some of the chapters and commented on them; I remain grateful to both for their help. A special word of thanks goes to Gary B. Madison, for a true philosophical friendship and for having introduced me to phenomenology and hermeneutics; there is little in this work that did not originate in my discussions with him. I am also grateful to Richard J. Bernstein, for sparking an interest in Gadamer's hermeneutics as well as American pragmatism, in the context of which my fascination with the horizon-problematic arose and developed. Just as grateful I am to Dmitri Nikulin, for showing me that nowadays, despite the demands of specialization, it is still fully possible to pursue a wide array of interests in philosophy in a serious fashion, by concentrating on problems and questions, rather than on fashions, schools, and traditions. I am especially grateful to James Dodd, my former Dissertation Advisor, for exemplifying how Husserlian phenomenology can be pursued as a living possibility as well as for providing me with the intellectual freedom to find my way into phenomenology.

This study is a rather heavily revised version of my dissertation, which I defended at the New School for Social Research a few days after the day that marked Husserl's birthday in 2008. I am grateful to all the members of my dissertation committee—James Dodd, Nicolas deWarren, Richard J Bernstein, and Dmitri Nikulin—for their support, advice, and encouragement. Let me also thank the Philosophy Department at the New School for Social Research for awarding my dissertation with the Hans Jonas Prize for the Best Dissertation in Philosophy, thereby giving me hope that this work might be of some interest to the larger philosophical community. A significant part of this manuscript was written at the Husserl-Archive in Cologne, where I spent a little more than a year in 2006–2007, thanks to a generous fellowship from the *Deutscher Akademischer Austauschdienst* (DAAD). I am grateful to Dieter Lohmar, who, whenever the opportunities would arise, always made it possible for me to return to the Archive and use the available resources to work on the revisions of this manuscript. I also owe a debt of thanks to James Madison University, for providing me with opportunities to return to the Husserl-Archive in Cologne in 2010 and 2011 to bring this manuscript to completion.

My daughter, Josephine, was born while I was working on this manuscript. Some parts of this work were composed with her in my arms. I am grateful to her for her laughter upon hearing me read difficult passages from Husserl aloud. Just as grateful I am to my parents and my sister, who, despite the geographical distance that separates us, continued to show interest in this work. Most of all, I am grateful to Geraldine.

Some chapters of this work, in their earlier versions, have already appeared in print. An earlier version of Chapter III, under the title "William James and Edmund Husserl on the Horizontality of Experience" was published in *Analecta Husserliana*, Vol. 108, ed. by A.-T. Tymieniecka, Dordrecht: Springer, 2011. Some of the themes that I address in Part III have appeared in print, although in a significantly abridged form, under the title "The Tremulous Grounds of Judgment: Husserl's Discovery of the World-Horizon," in *Urteil und Fehlurteil*, ed. by S. Loidolt and S. Lehmann, Wien-Berlin: Turia + Kant, 2011. I am grateful to the editors of these volumes for having shown an interest in the themes that are addressed in much greater detail in this work.

Contents

Abbreviations

Almost all of Husserl's works cited in this volume have been published in the *Husserliana* editions. Husserl's *Erfahrung und Urteil* as well as *Umsturz der kopernikanischen Lehre* constitute the only exceptions. Whenever possible, I have included the available translations of Husserl's works into English. When citing those volumes of *Husserliana*, which have not been translated into English yet, I have provided my own translations. Husserl's works that are published in Husserliana are cited as "Hua" + volume number + page number. Bibliographical references to these volumes can be found in Reference Lists, provided at the end of each chapter.

Hua I	*Cartesianische Meditationen und Pariser Vorträge*
CM	*Cartesian Meditations*
Hua III	*Ideen zu einer reinen Phänomenologie und phänomenologischen Philosophie. Erstes Buch: Allgemeine Einführung in die reine Phänomenologie* (ed. Walter Biemel, 1950)
Hua III/1	*Ideen zu einer reinen Phänomenologie und phänomenologischen Philosophie. Erstes Buch: Allgemeine Einführung in die reine Phänomenologie* (ed. Karl Schuhman, 1976)
Ideas I	*Ideas Pertaining to a Pure Phenomenology and to a Phenomenological Philosophy. First Book. General Introduction to a Pure Phenomenology*
Hua IV	*Ideen zu einer reinen Phänomenologie und phänomenologischen Philosophie. Zweites Buch: Phänomenologische Untersuchungen zur Konstitution*
Hua VI	*Die Krisis der europäischen Wissenschaften und die transzendentale Phänomenologie. Eine Einleitung in die phänomenologische Philosophie*
Crisis	*Crisis of European Sciences and Transcendental Phenomenology*
Hua VIII	*Erste Philosophie (1923–24). Zweiter Teil: Theorie der phänomenologischen Reduktion*
Hua XI	*Analysen zur passiven Synthesis. Aus Vorlesungs- und Forschungsmanuskripten (1918–1926)*
APS	*Analyses Concerning Passive and Active Synthesis. Lectures on Transcendental Logic*

Hua XIV	*Zur Phänomenologie der Intersubjektivität. Texte aus dem Nachlaß. Zweiter Teil (1921–1928)*
Hua XV	*Zur Phänomenologie der Intersubjektivität. Texte aus dem Nachlass. Dritter Teil (1929–35)*
Hua XVII	*Formale und transzendentale Logik. Versuch einer Kritik der logischen Vernunft*
FTL	*Formal and Transcendental Logic*
Hua XIX/1	*Logische Untersuchungen. Zweiter Band: Untersuchungen zur Phänomenolgoie und Theorie der Erkenntnis.*
LI	*Logical Investigations*
Hua XX/2	*Logische Untersuchungen. Ergänzungsband. Zweiter Teil. Texte für die Neufassung der VI. Untersuchung: Zur Phänomenologie der Ausdrucks und der Erkenntnis (1893/93-1921)*
Hua XXXIII	*Die 'Bernauer Manuskripte' über das Zeitbewußtsein (1917–1918)*
Hua XXXIV	*Zur Phänomenologischen Reduktion: Texte aus dem Nachlass (1926–1935)*
Hua XXXIX	*Die Lebenswelt. Auslegungen der vorgegebenen Welt und ihrer Konstitution. Texte aus dem Nachlass (1916–1937)*
Mat VIII	*Späte Texte über Zeitkonstitution (1929–1934). Die C-Manuskripte*
EU	*Erfahrung und Urteil. Untersuchungen zur Genealogie der Logik*
EJ	*Experience and Judgment: Investigations in a Genealogy of Logic*
"Umsturz"	"Umsturz der kopernikanischen Lehre in der gewöhnlichen weltanschaulichen Interpretation. Die Ur-Arche Erde bewegt sich nicht. Grundlegende Untersuchungen zum phänomenologischen Ursprung der Körperlichkeit, der Räumlichkeit der Natur im ersten naturwissenschaftlichen Sinne. Alles notwendige Anfangsuntersuchungen"
FI	"Foundational Investigations of the Phenomenological Origins of the Spatiality of Nature: The Originary Ark, the Earth, Does Not Move"

Chapter 1
Introduction

1.1 A Preliminary Determination of the Horizon

The following study aims to offer a systematic inquiry into the horizon-problematic in Husserl's phenomenology. A full-fledged determination of the notion of the horizon cannot be offered at the outset; its establishment must rest upon the subsequent investigation. Nonetheless, so as to provide the analysis itself with a direction, we are in need of the *guiding sense* of the theme under scrutiny. Since the philosophical notion of the horizon derives from the resources of everyday speech, the provisional determination of this notion can be developed on the basis of how the word "the horizon" figures in its everyday use.

I stand at the shore of the ocean, I look out into the distance, and I see the line at which the ocean meets the sky—such is the most common image one employs to characterize the horizon. From this common image, one can extract three dimensions of sense, and it is these dimensions that make up the preliminary sense of the horizon.

The horizon is the line at which the ocean or the sea meets the sky. The horizon thereby reveals itself as a *limit*, which constrains our visual field. Yet the limit in question need not be conceived as a negative force. After all, a phenomenon that falls outside the horizon is not a phenomenon at all. Moreover, not without a reason do we also speak of the horizons of *knowledge*, *experience*, and *interests*. Such a broad employment of this word indicates that the belongingness to the horizon as a necessary condition of manifestation should in no way be restricted to visual or even perceptual phenomena alone. The etymological sense of the horizon corroborates our expansive use of this term: the notion of the horizon derives from the Greek *ὁρίζειν*, a verb which can be roughly translated as "to divide," "to delimit," "mark off by boundaries," and which in Antiquity was employed primarily in the context of astronomy. Since delimitation is determination and since the meaning of any phenomenon rests upon its determination, the horizon reveals itself as the general framework, within which phenomena obtain their meaning. Such, then, is the first determination of the horizon: *the horizon is precisely what delimits phenomena; it is what allows phenomena to manifest themselves to us.*

S. Geniusas, *The Origins of the Horizon in Husserl's Phenomenology*,
Contributions to Phenomenology 67, DOI 10.1007/978-94-007-4644-2_1,

The horizon, as remarked, is a line that marks the intersection of the sea and the sky. It is, however, a peculiar line. Its peculiarity lies in the fact that *this line is neither drawn, nor can it be drawn.* If we were to draw it, we would already see what lies beyond it. And so, as Hegel, for instance, has shown in his critique of Kant, by positing a limit, consciousness immediately removes it.[1] Yet things appear different in regard to a line that in principle evades determination: *the more I approach it, the more it recedes; the more I advance toward it, the more it distances itself from me.* Thus already as an everyday term, the horizon reveals itself as *irreducibly malleable*. Due to its ineradicable plasticity, the horizon shows itself as an *unsurpassable* limit. Such being the case, no matter how much we can try to enlarge our field of vision, knowledge, or experience, the field in question nonetheless remains limited. This is the second sense of the horizon I wish to highlight: *the horizon is a limit that is in principle unsurpassable.*

Thirdly, even though the horizons always remain limited, nonetheless, they can always be modified and enlarged. We enlarge them by changing the situation we find ourselves in: by drawing nearer to the indeterminate line, or by moving away from it. The horizon thereby reveals itself as *relative* in regard to our whereabouts. This relativity gives rise to the realization that the general framework of sense that embraces phenomena is dependent upon our own situatedness. The question of how we came to inhabit our present standpoint thereby proves to be inseparable from the generation of the horizon itself. We thereby obtain the third sense of the horizon: *the horizon is relative in regard to our current situatedness.*

Such, then, is the threefold determination of the horizon: *the horizon is a versatile and unsurpassable limit, which is always relative to subjectivity and which delimits each and every phenomenon*. This threefold determination is crucial for the phenomenological appropriation of this everyday word. Yet it is just as crucial to emphasize that this determination of the horizon is only *preliminary*. Precisely because the phenomenological employment of this term is an attempt to expose, as Hans-Georg Gadamer would say, "the logical instinct of everyday speech,"[2] it cannot be an uncritical appropriation of how this notion figures in its common use. The provisional notion of the horizon needs to be supplemented with an account of the horizon as a philosophical notion.

1.2 The Horizon as a Philosophical Notion and Its Historical Origins

The *philosophical* application of the notion of the horizon can be traced back at least to the Neo-Platonic context of the doctrine of emanation.[3] We find the notion of the horizon already in the so-called *The Book of Causes* (*Liber de causis*),

[1] See, for instance, Hegel (1967, 110).

[2] See Gadamer's "Hegel und die antike Dialektik" (published in Gadamer 1971).

[3] For a more elaborate treatment of the themes I address in this section, see Scherner (1974), pp. 1188–1206.

a popular work in the Middle Ages, which was often falsely attributed to Aristotle. Although the authorship of this work still remains unknown, after William of Moerbeke translated Proclus' works into Latin, it became apparent that most of the content of this work was taken from Proclus' *Elements of Theology*. This fact allows one to claim that the Neo-Platonic appropriation of the notion of the horizon marks the emergence of this term in philosophy as a whole. In *Liber de causis*, the notion of the horizon emerges in the context of determining the place of the human soul, which is said to find itself in the horizon of eternity under and above time. However, even though the notion of the horizon emerges in the context of highly significant themes, in *Liber de causis* this notion occurs only a few times.

Although the Neo-Platonic problem of emanation by no means exhausts the employment of the notion of the horizon in Medieval philosophy, until the thirteenth century the task of determining the location of the soul with regard to time and eternity remained the overall context in which the philosophical application of the notion of the horizon found its home. The task of obtaining a better understanding of how the soul could find itself at the borderline between time and eternity remained the motivating force behind the employment of the notion of the horizon in the works of Isaak ben Salomon Israeli, Alanus of Lille, William of Auvergne, as well as Albertus Magnus. This context was also determinative of the manner in which the notion of the horizon figured in the works of Meister Eckhart. However, the situation changed with Thomas Aquinas, who was the first to give the notion of the horizon a specifically *anthropological* meaning. For Aquinas, it was no longer just the soul, but *the human being as a whole* that found itself within a specific horizon: being composed of the spiritual and physical nature, the human being was said to have the *limits* of both natures in itself.

One could thus say that in Medieval philosophy the notion of the horizon had both a *metaphysical* as well as an *anthropological* meaning. In this context, the horizon was for the most part conceived as a circle to which two hemispheres belong. The hemispheres in question were either those of time and eternity, or those of spiritual and material nature.

By contrast, in modernity the notion of the horizon obtains a specifically *epistemological* sense. The distinction between the true and the merely apparent horizons, conceived as the distinction between the individual and the universal horizons of knowledge, came to occupy a central place. Such an "epistemological turn" in the horizon-problematic meant that it was no longer possible to speak of the human being *finding itself* within a predetermined cosmological horizon. It was no longer a question of determining the place of the human soul, or the human being, within a metaphysical framework, which had been set in advance and which human beings could only subsequently identify, although by no means modify. As seen from the perspective of modernity, rather than *finding itself* in the predetermined ontological horizon, the human being *has* its own horizon of knowledge. The horizon thereby became a matter of *reflection* and *self-determination*.

Thus Gottfried Wilhelm Leibniz employs the notion of the horizon in the meditations on the possible breadth of human knowledge. Here the notion of the horizon plays an important role in determining the rational as well as the historical limits of human cognition. For Alexander Gottlieb Baumgarten also, the problematic of the horizon has to do with determining the limits of human knowledge. Baumgarten undertakes this task by drawing a distinction between those objects that can be known to us and the otherwise infinite universe of things. However, in contrast to Leibniz, Baumgarten's task is not so much that of determining the limits of human knowledge *in general*, but rather that of distinguishing between different capacities of cognition and determining the limits of these capacities. Thus in his *Aesthetica*, Baumgarten draws a distinction between sensuous knowledge (*cognitio sensitive*) on the one hand, and logico-rational knowledge on the other, and aims to determine the limits of these capacities.

The task of determining the breadth and limits of human knowledge remains the central task that underlies Immanuel Kant's employment of the notion of the horizon. Some have suggested that as a *philosopheme*, the horizon could be said to have its origins in Kant's works.[4] However, it is the *synthesizing powers* of Kant's philosophy that render his reflections on the horizon as remarkable as they are. Without much exaggeration, one could say that in Kant's writings, we face a synthesis of the different ways in which the notion of the horizon had been employed in modernity. On the one hand, Kant's interest in the horizon-problematic, much like Baumgarten's, is that of determining the limits of the different capacities of human cognition. In this regard, Kant draws a distinction between the logical, aesthetical and practical horizons, and thereby obtains more precise means to answer Baumgarten's central concern of distinguishing and delimiting the powers of human cognition. On the other hand, much like Leibniz, Kant is also interested in determining the breadth of human knowledge. By establishing a distinction between the questions the human mind can answer and those it cannot, Kant answers Leibniz's central concern as well. What is more, the distinction between the absolute and universal horizon on the one hand, and the private horizons on the other hand, also plays an important role in Kant's works. Kant thus speaks of the child's, the adult's, and family's horizons, of the horizon of philosophy, mathematics, and religion, of the horizons of opinions and of sciences.

Arguably, it is Kant's interest in the private horizons that has given rise to novel philosophical reflections on the horizon in the philosophy of the nineteenth century. Both in Friedrich Nietzsche's and Wilhelm Dilthey's writings, the notion of the horizon figures in the context of their reflections on history and culture. According to the position that Nietzsche defended in his *Use and Abuse of History*, it is the blindness to the actual historical horizons one inhabits that has given rise to the alarming consequences, which stem from the objectivism of modern historical science. As Hans-Georg Gadamer was later to observe, it was Nietzsche who first disclosed to us that the methodological orientation of historical sciences gives rise to a disturbing weakness of evaluation. As Gadamer put it, historical science

[4] See in this regard Kwan (1990).

> [h]as become so accustomed to considering things in ever different and changing lights that it is blinded and incapable of arriving at an opinion of its own regarding the objects it studies. It is unable to determine its own positions vis-à-vis what confronts it. Nietzsche traces the value-blindness of historical objectivism back to the conflict between the alienated historical world and the life-powers of the present. (Gadamer 1977, 7)[5]

In his own turn, Dilthey spoke of the horizon as the encompassing cohesion of a cultural system. According to Dilthey, each and every epoch is to be conceived as a different kind of a cultural system, and for this reason, each epoch brings with it a new horizon of sense, conceived as a limit that is set on our powers of sense and understanding. Put otherwise, each and every epoch finds its limits within the all-encompassing life-horizon.

This brief historical sketch makes it abundantly clear that in the history of philosophy, the notion of the horizon has received a number of not only complementary, but also conflicting determinations. As we have seen, the horizon has been conceived in a number of ways, the chief of which are its metaphysical, anthropological, epistemological, historical, and cultural determinations. It is by no means easy to reconcile the metaphysical and epistemological conceptions of the horizon, just as it is not clear how one is to unify the analysis of the private and the universal horizons. Yet even though the philosophical employment of the notion of the horizon reaches as far back as *Liber de causis*, nonetheless, I would question the conjecture one might be willing to make, viz., the conjecture that the horizon has been a *philosophical notion* starting with the Neo-Platonic employment of this term. Such might seem to be the case when the "history of the horizon" is addressed from the standpoint of an intellectual historian. However, so as to be identified as a *philosophical concept*, the horizon is at least in need of a clear definition. Arguably, in the long and diverse history of its philosophical employment, the notion of the horizon did not receive such a determination and figured as a *metaphorical* term, which would obtain new connotations of sense in different contexts of analysis. *In all these diverse contexts that I have just sketched, the notion of the horizon lacks a specifically philosophical determination.* In all of these contexts, the notion of the horizon remains a *metaphorical* term, which is employed in a cursory fashion without ever reaching a clear determination. As I will contend in the following investigation, it was Husserl who was the first, not only to transform the horizon into a notion of central philosophical importance, but even more notably, to uncover its full-blown problematic and significance. With this in mind, let us turn to the preliminary account of the horizon as a *phenomenological* notion.

---

[5] "Ich erinnere an seine [Nietzsches] Anklage der Wertungsschwäche, die den modernen Geist befallen hat, weil er sich so sehr gewöhnt habe, in immer wieder anderes und wechselndes Licht zu treten, so dass er geblendet sei und zu einer eigenen Wertung dessen, was sich ihm zeigt, zu einer Standortbestimmung gegenüber dem, was ihm gegenübertritt, nicht mehr fähig ist; die Wertblindheit des historischen Objektivismus wird hier zurückgeführt auf den Konflikt zwischen der verfremdeten geschichtlichen Welt und den Lebenskräften der Gegenwart" (Gadamer 1993, 221).

1.3 The Horizon as a Phenomenological Notion

In its widest sense, the *phenomenological* problematic of the horizon derives from what in Gestalt psychology is known as the *background/foreground schema*. When consciousness is thematically and intuitively aware of any object in the foreground, it is simultaneously aware, although non-thematically and non-intuitively, of the background as well. It would be insufficient to merely admit that this non-thematic and non-intuitive awareness always *accompanies* thematic and intuitive consciousness. Much more importantly, our awareness of the background always *determines* the manner in which we perceive the object in the foreground. This realization gives rise to a philosophically troubling question: *how can consciousness be aware of what is non-thematic and non-intuitive*? This question is particularly troubling for phenomenology—a philosophical tradition, which stands out from others due to the primacy it grants to *intuition*. Suffice it to recall the "principle of all principles," of which Husserl speaks in *Ideas I*: "*...everything originarily offered* to us *in 'intuition' is to be accepted simply as what it is presented as being*, but also *only within the limits in which it is presented there*" (*Ideas I*, 44).[6] With this principle in mind, one needs to ask: what sense is one to make of the awkward fact that our thematic intuition is itself dependent upon the intuitively empty consciousness of the horizon? Moreover, what does this dependence tell us about consciousness itself and, more generally, about subjectivity?

The phenomenological analysis of the horizon aims to provide an answer to these far-reaching questions. In the present context, let me sketch the phenomenological answer in a rough-and-ready form: the phenomenological answer suggests that the non-thematic and non-intuitive consciousness is possible on the basis of subjectivity's *prior experience* that consciousness, due to its temporal structure, manages to retain. The capacity of retention gives rise to the *sedimentations* of experience, which in their own turn make consciousness capable of "horizonal protentions," i.e., they enable consciousness to project meaning into what is directly perceived.[7]

Yet why would phenomenology privilege *experience* in the context of the horizon-problematic? Allow me to turn to a rather lengthy illustration to clarify this issue. When woken up suddenly from a vivid dream, one can find oneself at a loss in regard to the time and the location of one's whereabouts. Now-and-again the content of one's dream intrudes into the state of wakefulness, thereby giving rise to an awkward congruence: the world of dreams thrusts itself upon one's actual appearances and

[6] "Am *Prinzip aller Prinzipien*: daß *jede originär gebende Anschauung eine Rechtsquelle der Erkenntnis sei, daß alles*, was sich uns *in der 'Intuition' originär*, (sozusagen in seiner leibhaften Wirklichkeit) *darbietet, einfach hinzunehmen sei, als was es sich gibt*, aber auch *nur in den Schranken, in denen es sich da gibt*, kann uns keine erdenkliche Theorie irre machen." (Hua III/1, 51)

[7] Let me emphasize repeatedly what I have already mentioned in Sect. 1.1: the phenomenological thematization of the horizon is not restricted to perceptual phenomena. Rather, the background / foreground schema constitutes a *model* according to which *any possible object* lends itself to understanding.

envelops them within its own edifice. Under such conditions, the dim light that reaches one from another room, or the rattling of the light rain that gently obtrudes through the open window, bewitches one as the perseverance of one's dream. In such situations, the figures and scenes of the dream coalesce with the objects that dimly fall into one's perceptual field, be it a wardrobe, a bookcase, or their shadows. Yet one is not held captive by one's dreams too long. Just as we are all aware of such strange fusions of one's self with the content of one's dreams, so we are also aware how still in darkness, the world of dreams separates itself from the world of wakefulness: the furniture regains its usual face, the light rain and dim light are recognized for what they are, thereby forcing the content of the dream to recede into the background, from which this content will never spring up again within the thematic field.

It is by no means a prerogative of vivid *dreams* to carry into effect such metamorphoses of concrete objectivities. One can just as much be overpowered by a dazzling *memory* or a resplendent *phantasy* in such a way that whatever emerges in the perceptual field is *immediately* apperceived as a figure that belongs to the world of memory or phantasy. I emphasize the word "immediately" because, as one reflects on such metamorphoses more cautiously, one realizes that it is not full-fledged objectivities that undergo such curious modifications. Clearly, one does not recognize the rattling of the rain or the beam of light for what they truly are before transforming them into memorial or phantasy-images. Rather—and this is the point the illustration is meant to bring forth—it is an *appearance still devoid of perceptual sense* that comes to be *immediately absorbed* within a particular horizon of sense, from which it derives its determinacy.

What comes to the fore with such a realization no longer pertains to the specific types of consciousness addressed thus far. Rather, this *immediate absorption* bespeaks an *essential characteristic of consciousness itself*: whatever world or quasi-world it inhabits, be it the world of dreams, of memory, of phantasy, or of perception, *consciousness always absorbs an appearance within a certain horizon of sense*, dresses it up, as it were, with a specific panoply of meaning, due to which an appearance manifests itself as an appearance of a particular objectivity. Only due to such powers of absorption can consciousness be faced with full-fledged objectivities and, more generally, can consciousness inhabit a meaningful world.

The notion of the horizon is the word Husserl employs to indicate, describe, and elucidate such *powers of absorption*, without which consciousness could not be aware of full-fledged objectivities. In its broadest determination, the horizon, as a philosophical notion, can be characterized as *what consciousness co-intends in such a way that what is co-intended determines the sense of appearing objectivities*. One could say that the horizon embraces what one is conscious of in a non-thematic way. Itself remaining non-thematic, the horizon keeps the space open for the emergence of each and every theme. The horizon is a structure of determination that predelineates the purview within which each and every phenomenon appears. The horizon belongs to appearances, yet it does not (at least directly) derive from appearances. Rather, it is the intuitive emptiness, which is nonetheless given, and given as inseparable from the intuitive fullness. It is a peculiar self-givenness of consciousness, which proves to be indispensable for the self-givenness of appearances as objectivities. It is a more or less general framework of sense that embraces appearing phenomena. As seen from the perspective of everyday-experience, the horizon is a pregiven origin of sense from which appearances derive their determinacy.

Yet where does this "origin of sense" itself derive from? My earlier remark that the horizon does not *directly* derive from appearances does not exclude the possibility that in a more roundabout way, appearances might themselves constitute the origin of the horizon. As we will see, the so-called genetic phenomenology defends precisely such a hypothesis. One of the central goals of the following analysis is to show that an inquiry into the origins of the horizonal structures of experience holds the promise of significantly enriching our understanding of subjectivity.

1.4 The Question of Origins

Although by now the general sense of the horizon-problematic has been brought to the fore, the meaning of *the question of origins* remains indeterminate. What could it possibly mean to raise the question of the *origins* of the horizon? This question—and thus the title of this study as well—lends itself to a *threefold* interpretation.

First, the horizon is a *philosophical* notion *and a philosophical* theme. In this regard, to raise the question of the origins of the horizon is to ask: how did the everyday word "the horizon" become a problem for philosophy? One of the tasks of the present study is to show that Husserl is the philosopher to whom we are to this day indebted for having transformed the metaphorical employment of this term in the history of philosophy into a specifically *philosophical* notion and a specifically *philosophical* theme.

Yet if one understood the question of the origins of the horizon in this sense alone, its significance would be severely limited. Under such circumstances, one could even argue that the question of origins is not a philosophical question at all, that it is of importance only to those interested in a forgotten chapter in the recent history of ideas, and to this one could further add that even in such a restrictive context, this forgotten chapter is only of marginal significance. It is therefore crucial not to overlook that the question of the origins of the horizon is of much greater importance than its first sense suggests.

Thus secondly, let us ask: what exactly does the philosophical notion of the horizon aim to depict? One of the central answers that stems from Husserl's phenomenology suggests that *the horizon is a peculiar structure of experience*. As I have already indicated in the previous section, the horizon depicts the capacity of consciousness to enwrap appearances within the structure of familiarity. Once so much is said, one further wonders: what could it mean to raise the question of the origins of the horizon in such a framework? In light of this concern, the question of origins obtains a new sense, where it means nothing other than an attempt to clarify the *genetic set-up of experience itself*. One can thematize this question by raising the following concerns: how does subjectivity obtain the above-mentioned structures of familiarity?

Don't these structures derive from experience? Moreover, do they not undergo continuous modifications and transformations? Such, then, is the second sense in which the question of the *origins* of the horizon lends itself to understanding: to raise this question is to inquire into the crystallization of the structures of experience. In short, this second sense suggests that the question of the origins of the horizon is the question of the origins of the horizon of *subjectivity*.

Yet an inquiry into the problematic of the horizon would still remain incomplete even when the question of the origins of the horizon is understood in these two distinct senses. After all, doesn't Husserl himself speak of the *world-horizon*, and does he not identify the world as the most basic and irreducible figure of the horizon? Furthermore, does Husserl not himself suggest that the reason why the world-horizon remained unthematized in many of his early writings had to do with an exclusive preoccupation with the horizons of subjectivity?[8] With this in mind, let me supplement my initial portrayal of the horizon with the following observation: in Husserl's phenomenology, *the world is claimed to be the rudimentary figure of the horizon*. But what could it mean to raise the question of the origins of the horizon in such a framework? It would mean nothing other than to inquire into the different senses of the *world's pregivenness*.

Thus Husserl's frequent identification of the world as the originary figure of the horizon indicates the third sense in which the horizon-problematic proves to be inseparable from that of origins. This conjunction not only bespeaks the phenomenological solicitation to acknowledge the world's pregivenness in regard to all appearing phenomena; more importantly, it also places a demand to inquire into the formation of the world's pregivenness, a demand to thematize the genesis of world-experience itself.

In what follows, I will argue that the most striking feature that differentiates Husserl's inquiry into the horizon from the development of this theme in post-Husserlian thought pertains to the circumstance that *in Husserl's phenomenology, the horizon-problematic unfolds as inseparably tied to the question of origins*. According to my central thesis, *the horizon is nothing other than the conceptual space, which legitimates the phenomenological inquiry into the origins of sense-formation*. What is more, these sense-formations can be investigated in two different ways: either within the context of the *horizon of subjectivity* or within the context of the *world-horizon*. As will become apparent in the conclusion to this investigation, this two-sidedness of the horizon-problematic is by no means accidental in that it derives from the fact that the horizon is a peculiar figure of *intentionality*. Such a two-sided inquiry into the origins of the horizon will prove to be an inquiry into the formation of habitualities of consciousness, i.e., into the crystallization of experience itself.

Husserl is the *first* philosopher to have raised the question of the origins of the horizon *explicitly*, and in the context of the horizon-problematic, he granted this question preeminent significance. Yet surprisingly, Husserl is also the *last* philosopher

[8] See, for instance, Hua III, Beilage XIII, pp. 399–400.

to have problematized the horizon by raising such a concern. As we will see, the question of origins has almost completely disappeared from the landscape of philosophy in post-Husserlian thought. Starting with the second-generation phenomenologists and hermeneuticists, this question has, on a number of fronts, been interpreted as an *illegitimate* question, which severely constrains Husserl's concrete investigations.

These popular critiques notwithstanding, one of the central tasks of this study is that of reviving the question of the origins of the horizon as a *necessary question* for the philosophical understanding of the horizon-problematic. It is my contention that in the absence of the question of origins, our understanding of the horizon would remain inadequate; it is also my contention that in the absence of the horizon-problematic, our understanding of subjectivity would remain incomplete and distorted.

1.5 The Structure of the Following Investigation

Husserl is the most prolific philosopher of the twentieth-century and the notion of the horizon is one of his most commonly used terms. No one could possible go through all of Husserl's published and unpublished manuscripts in detail, let alone provide a detailed commentary on them. Thus any attempt to provide an exhaustive account of the horizon-problematic can only be preliminary. *An investigation of the horizon can only be horizonal.*

When choosing to address certain themes at the expense of others, I have been guided by *the question of origins.* This choice does not only rely on the fact that the question of origins is privileged in Husserl's writings; no less importantly, by pursuing this question, one can bring to light the *unique* contribution to the horizon-problematic that stems from the Husserlian framework.

Thus the threefold sense in which the question of the origins of the horizon lends itself to interpretation (see Sect. 1.4) will guide the following investigation. Accordingly, this study is divided into three parts. Part I of this work (Chaps. 2, 3, 4 and 5), will address the emergence of the horizon in Husserl's phenomenology. Part II (Chaps. 6, 7, 8 and 9) will thematize the origins of the horizon as the formation of the *horizon of subjectivity*. The third part (Chaps. 10, 11 and 12) will turn to the *world-horizon* as a rudimentary figure of the horizon.

1.5.1 Part I: The Emergence of the Horizon

As my foregoing reflections suggest, to raise the question of the emergence of the horizon in Husserl's phenomenology is nothing other than to inquire into how this notion became a genuine philosophical concept. It is, of course, all-too-easy to get lost in the thousands of pages of Husserl's unpublished manuscripts. One thus wonders how the origins of the horizon in philosophy could be fixed more precisely.

Fortunately, Husserl himself provides us with a helpful clue. As he remarks in *Formal and Transcendental Logic*, "In the *Logical Investigations* I still lacked the theory of horizon-intentionality, the all-determining role of which was first brought out in the *Ideas*" (FTL, 199).[9] This remark will continue to guide my analysis in the next four chapters of this work. In unusual terseness and simplicity, this remark seems to fix the origins of the horizon in a precise way: it singles out *Ideas I* as a text whose publication announces the inauguration of the horizon-problematic in philosophy. When it comes to the search for the origins of the horizon, one's path thus *seems* to be delineated: apparently, it remains to follow the analysis undertaken in *Ideas I* and on the basis of this analysis, to tell a story about the origins of the horizon.

The few interpreters who have addressed the question of the emergence of the horizon have followed precisely this path.[10] Nonetheless, *this path is too straightforward and, at the end of the day, it leads one astray.* Admittedly, *Ideas I* is the text in which the *notion* of the horizon originates; yet the *problematic* of the horizon precedes the publication of this work.

In *Formal and Transcendental Logic*, after identifying *Ideas I* as the work in which the notion of the horizon originates, Husserl continues in the following way: "therefore, in the earlier work, I could not finish with occasional judgments and their signification" (FTL, 199).[11] Thus even the passage from *Formal and Transcendental Logic* that I have just quoted—the passage that *seems* to invite one to begin the inquiry into the emergence of the horizon with *Ideas I*—in fact motivates one to start such an investigation by turning to Husserl's earlier works. A closer look at the passage motivates one to begin with five questions.

First of all, one needs to ask: why is the notion of the horizon of such great importance for the problem of *occasional expressions*? Surprisingly, this question is overlooked in the works that have been dedicated to the emergence of the horizon, *even though the assurance that Ideas I signals the emergence of the horizon in philosophy stems from the above quoted passage*. So as to counteract this shortcoming, in what follows, I will instigate the analysis of the horizon by turning to occasional expressions as they emerge in the first of the *Logical Investigation* (1900–1901).

Secondly, what sense is one to make of this passage alongside Husserl's other remarks, which suggest that the horizon-problematic originates in William James' *Principles of Psychology*?[12] Without any doubt, our understanding of the horizon would remain imprecise for as long as we do not inquire into the relation between Husserl's phenomenology and James' psychology. I will turn to this issue in Chap. 3.

---

[9] "In den *Logischen Untersuchungen* fehlte mir noch die Lehre von der Horizont-Intentionalität, deren allbestimmende Rolle erst die *Ideen* herausgestellt haben" (Hua XVII, 177).

[10] See Kuhn (1940) and Kwan (1990).

[11] "Darum konnte ich dort mit den okkasionellen Urteilen und ihrer Bedeutung nicht fertig werden." (Hua XVII, 177).

[12] See, for instance, Hua VI, 267 and Cairns (1976, 36).

Thirdly, in *Ideas I* Husserl employs the terms *Horizont*, *Hof* (halo), and *Hintergrund* (background) interchangeably. For instance, in §83 Husserl remarks: "'horizon' thus has the same value here as the terms 'halo' and 'background'" (*Ideas I*, 197).[13] This claim is also problematic, because even though prior to *Ideas I* Husserl does not employ the term *Horizont*, he employs the terms *Hof* and *Hintergrund* as early as the *Logical Investigations*. The inquiry into the emergence of the horizon must therefore clarify how one can distinguish between *Horizont* on the one hand, and *Hintergrund* and *Hof* on the other, so as to extract the unique significance of the former notion. I will address this question still in Chap. 3.

Fourthly, in the quoted fragment Husserl speaks of the all-determining role (*allbestimmende Rolle*) of horizon-intentionality. This is also deeply problematic because, as Husserl's other texts make clear (and as will become plain in Chaps. 4 and 5), *Ideas I* does not generate a full-fledged determination of the horizon. One is thereby led into an impasse: how can one both affirm and deny the all-determining role of the horizon in *Ideas I*? I will turn to this question in Chap. 4.

Finally, on a closely related point, how is one to understand the curious situation that Husserl in *Formal and Transcendental Logic* (1929), looking back at *Ideas I* (1913), identifies the discovery of the horizon as one of this work's fundamental accomplishments, *even though*, as we will soon see, *in the text itself the notion of the horizon appears only in passing*? How can Husserl see in this text something that it itself, in the strict sense, does not possess? I will bring my analysis in Part I to completion by engaging this question in some detail.

The four chapters that fall into the first part of this work are meant to deal with the outlined set of issues. In Chap. 2, I will raise the question of indexicality's *phenomenological* sense by tracing the development of this problem in Husserl's phenomenology, starting with its emergence in the *Logical Investigations*. In contrast to the standard approach, which confines the problem of indexicality to its treatment in the *Logical Investigations*, I will argue against Husserl's early solution. I will suggest that, from a specifically phenomenological perspective, the so-called "replaceability thesis" is unwarranted. I will further show that Husserl himself has unequivocally rejected his early solution in his revisions of the *Logical Investigations*, although admittedly, he has never replaced his old conception with a new one. My central task in this chapter will be that of *reconstructing* the main contours of Husserl's new approach to indexicality. Following Husserl's own suggestion that the discovery of the *horizon* puts phenomenology in the position to rethink the problem of the indexicals, this chapter will procure a preliminary understanding of horizon-intentionality in Husserl's writings and show how the dynamic structure of the horizon invites the question of the genesis of expressibility. In the final analysis, the presence of the indexicals in the scientific discourse will turn out to be a faint echo of the life-world from which scientific discourse springs.

[13] "Horizont gilt hier also soviel wie [...] die Rede von einem 'Hof' und 'Hintergrund.'" (Hua III/1, 186, Anm. 1)

Chapter 3 is primarily designed to answer the following question: in which sense is the horizon a specifically *philosophical* notion and a specifically *philosophical* theme? With this in mind, the chapter will examine how Husserl's notion of the horizon derives from James' analysis of the halo, background, and fringe of consciousness. I will argue that Husserl is to be considered the founder of the horizon-problematic in philosophy, yet not because he was the first to have thematized the phenomenon of the horizon. As we will see, James had already done so at a great depth. The significance of Husserl's analyses rather consists in having *depsychologized* the problematic of the fringe of consciousness and in having disclosed its *transcendental* dimensions. Thus, as a philosophical notion and a philosophical theme, the horizon will turn out to be irreducibly *transcendental*.

Chapters 4 and 5 will inquire into Husserl's thematization of the horizon in *Ideas I*—the work that marks the emergence of the horizon-problematic in Husserl's phenomenology. In Chap. 4, I will turn to the *world-horizon*, which, from the historical point of view, is the originary figure of the horizon in Husserl's phenomenology. However, my analysis will lead to a somewhat paradoxical realization that it is precisely the world-horizon that remains unexplored in *Ideas I*. On this basis I will argue that in the work that marks the emergence of the horizon-problematic in philosophy, the horizon in its all-determining sense is both uncovered and suppressed, revealed, yet not effectively resolved. Put otherwise, *Ideas I* procures only a preliminary and not a conclusive notion of the horizon.

Chapter 5 will turn from the problematic of the world-horizon to that of the horizons of concrete objectivities. My analysis will show that the horizon is a distinctly *transcendental* notion, which qualifies *intentional* consciousness and which can be described phenomenologically on the basis of the *epochē* and the reductions. By linking the problematic of the horizon to the consciousness of the "I can" and the different figures of givenness in Husserl's phenomenology, I will argue that the horizons are not to be thought of as characteristics of nature and natural things, but as indispensable components of lived-experiences and objects constituted in these experiences. Having shown how the horizon-problematic entails *three different* forms of self-consciousness, my analysis will lead to the conclusion that, despite the richness of Husserl's early account of the horizon, the analysis provided in *Ideas I* remains deficient, and for two reasons: it fails to inquire into the horizons of transcendental subjectivity just as it fails to uncover the world-horizon. By identifying the "Cartesian way to the reduction" as the methodological reason that underlies this twofold limitation, I will suggest that the transition from static to genetic phenomenology is up to a large degree motivated by the need to broaden the horizon-problematic. My analysis will disclose the horizon as essentially two-sided, embracing not only the objects of lived-experience (*Erlebnis*), but also lived-experience itself. Yet we will also be led to recognize that in *Ideas I*, Husserl had not carried through his inquiry into the horizon's correlative structure to its end. Husserl's early analysis of the horizon will thus appear similar to the painter's first sketch on the canvas: cursory, hasty, in short, premature.

1.5.2 Part II: The Horizons of Transcendental Subjectivity

Since the analysis undertaken in Part I of this work culminates in the realization that Husserl's incipient investigations of horizon-intentionality generate questions to which they fail to offer satisfying answers, the task of Part II is that of following these questions as clues so as to demonstrate how a direct engagement in them reveals the horizon in its genetic phenomenality. While I will argue that the horizon is a specifically genetic phenomenon, I will also suggest that the difference between Husserl's static and genetic accounts of horizon-consciousness cannot be grasped if one assumes from the start that the genetic thematization of the horizon signifies a complete abandonment of the static account and thus is a drastic alternative to those dimensions of sense of horizon-intentionality which have already been disclosed in the first part of this work. Rather, this transition should be seen as a progressive deepening of those intuitions, which at times are partly disclosed, and at other times only hinted at, in *Ideas I*.

In Chap. 6, I will argue that the horizon is a specifically *genetic* theme, which in its early emergence is still dressed in static garb. So as to establish this thesis, I will address the difference between static and genetic phenomenological methods and further inquire into the difference between the static and genetic notions of horizon-consciousness. Such a twofold determination of consciousness will be further supplemented with a static and a genetic determination of the ego. On such a basis, I will further show how the genetic notion of the ego procures a phenomenological notion of *transcendental subjectivity*. I will argue that transcendental subjectivity is one of the greatest discoveries of genetic phenomenology (the other being that of the world-horizon) and that the task of the genetic analysis of the horizon is first and foremost that of uncovering the horizons of transcendental subjectivity.

In Chap. 7, I will argue that phenomenology cannot disclose the horizons of transcendental subjectivity for as long as it understands the phenomenological reduction in accordance with how it was spelled out in the confines of static phenomenology. I will argue that the philosophical significance of the *new* path to the reduction, which Husserl has introduced in *First Philosophy II*, lies in the fact that it enables phenomenology to disclose the distinct horizons of transcendental subjectivity. I will suggest that from *First Philosophy II*, one can derive a new notion of the horizon, conceived as the milieu of concealed sense accomplishments, i.e., as the very conceptual space that genetic phenomenology aims to thematize. On such a basis, the question of the origins of the horizon will obtain its specifically phenomenological sense: it will merge with the question of sense-formation itself.

In Chap. 8, I will argue that *both the sense and the methodological justification of the phenomenological question of origins derive from the problematic of the horizon*. I will show that Husserl's notion of the horizon entails two dimensions of sense: the horizon is a horizon of *reference* (*Verweisungshorizont*) and of *validity* (*Geltungshorizont*). As a *system of reference*, the horizon embraces all the implications that each appearance draws to other appearances. The qualification of the horizon as a *system of validity* entails a further realization that an actual appearance entails references not only to other actual appearances, but also to other *potential* modes of appearances. I will argue that the phenomenological question of origins is first and foremost meant to trace the concealed sense-accomplishments, which qualify the sense of any appearing objectivity. On the basis of what is stated above, I will suggest that (1) the horizon as a system of validity clarifies *the sense* of the question of origins, and that (2) the *possibility* of the question of origins is secured by the horizon as a system of reference.

In Chap. 9, I will bring Husserl's genetic notion of the horizon into a dialogue and confrontation with the hermeneutical accounts of the horizon. I will suggest that the hermeneutical critique of Husserl's phenomenology provides a new impetus to disclose the philosophical significance of the *historicity of transcendental subjectivity*. With this in mind, I will situate Husserl's notion of the primal ego in between the hermeneutical critique and the "functional reading" of this concept. I will trace the development of the notion of the primal ego by interpreting this concept in the framework of the *Bernau Manuscripts*, *C-Manuscripts,* and the *Crisis*. I will argue that the primal ego is not to be thought of as an entity, but rather as a notion that indicates a number of different levels of transcendental subjectivity's constitutive accomplishments. On such a basis, I will argue that neither the hermeneutical critique nor the functional interpretation exhausts the significance of Husserl's notion of the primal ego. Rather, the primal ego is a notion, which is first and foremost designed to indicate the historicity of transcendental subjectivity.

1.5.3 Part III: The World-Horizon as the Wherefrom, Wherein, and the Whereto of Experience

My analysis of the horizon as the conceptual space, which legitimizes the question of origins, will culminate in the realization that the world-horizon is the ultimate origin from which all sense-formations spring. Yet is Husserl's concept of the world-horizon robust enough to capture and disclose the phenomenality of the world? As we will see, some of Husserl's closest students and most allegiant followers were led to answer this question with an uncompromising *no*. The central goal of the third part of this study is to show why such a dismissive view is indefensible. That is, the goal is to show how Husserl's inquiry into the givenness of the world uncovers dimensions of sense that are inseparable from the world's phenomenality.

According to the central thesis that I will aim to substantiate in Part III, the concept of the world-horizon is meant to disclose the world as the *wherefrom*, the *wherein*, and the *whereto* of the appearance of concrete objectivities. As the *wherefrom*, the world shows itself as the *ground* (*Boden*) of experience; as the *wherein*, it reveals itself as an *absolute substrate* and as the *field of fields*, thereby accounting for the unity of experience; as the *whereto*, it presents itself in a *continuous process of constitution*, thereby circumscribing the *productivity of experience*.

In Chap. 10, I will argue that (1) Husserl's notion of the *world-horizon* is to be conceived as the ultimate origin from which all sense-formations spring, and that

(2) the world-horizon, conceived as the *wherefrom* of experience, constitutes the most original figure of the world-horizon. I will argue that in its most rudimentary manifestation, the world-horizon is given as *non-objective*, *non-thematic*, and *non-intuitive*, yet despite such a threefold negative qualification, one is in full right to speak of the *givenness* of the world-horizon. This realization will necessitate one to give up the conjecture one might be willing to make, viz., the conjecture that the horizon as a phenomenological concept can be conceived on the basis of the background/foreground schema.

Chapter 11 will turn to the second determination of the world-horizon, conceived as the *whereto* of experience. As we will see, the characterization of any entity as *in-etwas* calls for a correlative qualification of the world as *All-etwas*, and such a characterization of the world constitutes the basis of the world-horizon as the *wherein* of experience. I will subject such a notion of the world-horizon to Jean Wahl's critique, according to which phenomenology lacks the conceptual means to assert that *totality* and *possibility* constitute two aspects of the world-horizon. I will answer this objection on the basis of the distinction Husserl draws between *open* and *problematic* possibilities. In the final analysis, the world-horizon as the *wherein* of experience will have to be conceived on the basis of *open* possibility.

Chapter 12 will address the third and final figure of the world-horizon: the world-horizon as the *whereto* of experience. A retrospective glance at the first two figures of the world-horizon will bring to light a common presupposition: whether one conceives the world-horizon as the *wherefrom* or the *wherein* of experience, one limits the world-horizon to what is given through *sensuous* and *unmodalized* experience. By contrast, to thematize the world-horizon as the *whereto* of experience is to describe in its essential features the enrichment of sense that the world-horizon undergoes due to *modalization* and *non-sensuous experience*. This chapter will show how the world-horizon, as the *whereto* of experience, is linked to the productivity of modalized experience, the historicity of subjectivity, the continuous generation of new types of experience, and finally, the streaming-in of subjective accomplishments into world-experience. The chapter will also address the relation between the concepts of the world-horizon and the life-world.

Given such a multifaceted way in which the horizon lends itself to interpretation in the context of Husserl's phenomenology, we will also have to ask whether the horizon could be said to be a *notion* at all, i.e., whether it is not just a semblance of numerous themes that are only by chance given one and the same name. This study will end with an inquiry into this question. Thus in the Conclusion, I will turn to the question of the unity of the horizon. I will argue that the horizon obtains its unity as a *figure of intentionality*. By turning to Merleau-Ponty's and Derrida's readings of Husserl's "Origin of Geometry," I will suggest that the horizon is not just accidentally, but rather necessarily two-sided: it is both the *horizon of subjectivity* and the *world-horizon*. One could say without any exaggeration that this two-sidedness is the horizon's "universal apriori of correlation." I will conclude by suggesting that the strength of Husserl's way of thematizing the horizon consists in disclosing subjectivity in terms of those dimensions, which remain overlooked in post-Husserlian approaches to the horizon-problematic, especially as they are exemplified in philosophical hermeneutics and in French phenomenology.

1.6 The Indeterminacy of the Horizon in Husserl's Phenomenology and in Post-Husserlian Thought

My foregoing remarks should have clarified both the title and the structure of the following study. Yet one might wonder, does the question of the origins of the horizon need to be posed at all? What is it that motivates such an investigation? What is it that makes it, in any loose sense, "necessary?"

With these questions in mind, I would like to draw attention to three circumstances. (1) One should not overlook that despite its central significance, the notion of the horizon remains indeterminate in Husserl's works. (2) Nor should one fail to notice that the horizon-problematic remains largely overlooked and unexplored in the commentaries on Husserl's phenomenology. (3) Last but not least, the horizon-problematic cannot be restricted to Husserl's phenomenology alone. After all, the notion of the horizon does not just appear in post-Husserlian philosophy; it plays an important role in a number of other philosophical traditions. And yet, the notion of the horizon emerges in diverse philosophical contexts as a vague notion and a metaphorical theme, whose importance equals its opacity. Let us take a closer look at these three circumstances.

1. Borrowing a phrase from Eugen Fink,[14] one could characterize the horizon as an *operative* notion. By this I mean that the horizon determines a large variety of other phenomenological themes while it itself remains without definition. This lack of determinacy stands in sharp contrast to the horizon's central significance in Husserl's phenomenology. Paradoxically, this lack of determinacy does not derive from the horizon's under-determination, but on the contrary, from its *over-determination*. Starting with its discovery in *Ideas I* (1913), Husserl progressively employs the notion of the horizon with growing frequency, sometimes even inserting it into his revisions of earlier works.[15] Husserl's use of this term on almost each and every page of some of his writings signifies a progressive enrichment of its sense, and yet, it simultaneously signifies its accelerating obscurity. We still lack a systematic and comprehensive account of this central phenomenological theme.
2. Such an account is further called for by the realization that the notion of the horizon remains almost as indeterminate in the critical literature on Husserl's phenomenology as it is in Husserl's works themselves. Instead of being approached as a fundamental phenomenological theme that envelops Husserl's more detailed analyses, for the most part, the horizon-problematic attracts the attention of phenomenologists as a stratum within the context of other, supposedly more significant, thematic concerns. Given Husserl's frequent and indeterminate employment of the notion of the horizon, such a state of affairs becomes understandable, yet by no means justifiable.

---

[14] See Fink (1976), pp. 180–205.

[15] See, for instance, Husserl's revisions of his early analyses of time-consciousness (Hua X) or his lectures from 1910–11, *Grundprobleme der Phänomenologie* (published in Hua XIII).

There are, however, notable exceptions. One should not overlook the articles written by Helmut Kuhn[16] and Kwan Tze-Wan,[17] as well as more extensive studies undertaken by Donn Welton[18] and especially by Roberto Walton.[19] These investigations illuminate the different senses that the notion of the horizon embraces as well as provide a broad overview of the development of this theme in Husserl's works. They thereby generate an impetus to inquire into the origins of the horizon in Husserl's phenomenology.

3. The horizon is a philosophical theme that does not exclusively belong to Husserl's phenomenology. Yet curiously, having been appropriated by a number of post-Husserlian thinkers, the horizon emerges in these different philosophical contexts just as metaphorically as it did in Husserl. In Merleau-Ponty's early studies, the problematic of the horizon emerges in the context of the phenomenological discussion of the lived-body, and it leads toward the disclosure of its anonymous horizons and its anonymous life.[20] With the discovery of the notion of the *flesh* (*chair*) in his late works, Merleau-Ponty further thematizes the horizon as a "polymorphic matrix" and a "new type of being."[21] In Heidegger, we encounter the notion of the horizon in *Being and Time*. After the publication of this work, Heidegger rarely employs this notion, arguably due to the epistemological connotations that it entails. Yet in *Being and Time*, the phenomenological analysis of time is an inquiry into the "transcendental horizon of the question of being." In this text, the "horizon of being" is disclosed and interpreted as transcendence seen in its originary ecstatic emergence. In Gadamer, we witness the emergence of the horizon as a central hermeneutical theme in the framework of the problem concerning the relation between the universal and the particular. In more recent French philosophy, we find a recurrent attempt to break free from the boundaries of the horizon.[22] Yet interestingly enough, none of these attempts to deepen Husserl's notion of the horizon, or to abandon it, provides a determination of the horizon; for the most part, even a definition of the horizon is missing.

The task of the following investigation is not that of providing a detailed account of the role that the notion of the horizon plays in these diverse philosophical contexts. My goal is more modest and the references to these diverse contexts serve the purpose of showing that the question of the origins of the horizon is by no means of significance only for Husserlian phenomenology. So as to recognize the novelty and evaluate the significance of the alternative accounts of the horizon, one needs to acknowledge the resources from which they spring and against which they take their departure. The present investigation is directed at these resources.

[16] See Kuhn (1940).

[17] See Kwan (1990).

[18] See Welton (1983, 2000, 2003).

[19] See Walton (1991, 1997, 2002, 2003).

[20] See Merleau-Ponty (1976, 94, 197).

[21] See Merleau-Ponty (1968, 195, 281, 300, 339).

[22] See Marion (1998, 2002).

References

Cairns, Dorion. 1976. *Conversations with Husserl and Fink*. The Hague: Martinus Nijhoff.

Fink, Eugen. 1976. *Nähe und Distanz: Phänomenologische Vorträge und Aufsätze*. Freiburg/München: Verlag Karl Alber.

Gadamer, Hans-Georg. 1971. *Hegels Dialektik: Fünf hermeneutische Studien*. Tübingen: J.C.B. Mohr (Paul Siebeck).

Gadamer, Hans-Georg. 1977. The universality of the hermeneutical problem. In *Philosophical hermeneutics*. Trans. D. Linge. Berkeley: University of California Press.

Gadamer, Hans-Georg. 1993. Die Universalität des hermeneutischen Problems. In *Gesammelte Werke*, Band II. Tubingen: J.C.B. Mohr (Paul Siebeck).

Hegel, Georg Wilhelm. 1967. *Wissenschaft der Logik. Erster Teil*. Leipzig: Meiner Verlag.

Husserl, Edmund. 1950. In *Ideen zu einer reinen Phänomenologie und phänomenologischen Philosophie. Erstes Buch: Allgemeine Einführung in die reine Phänomenologie*, ed. Walter Biemel. The Hague: Martinus Nijhoff.

Husserl, Edmund. 1966. In *Zur Phänomenologie des inneren Zeitbewusstseins (1893–1917)*, ed. Rudolf Boehm. The Hague: Martinus Nijhoff.

Husserl, Edmund. 1969. *Formal and transcendental logic*. Trans. Dorion Cairns. The Hague: Nijhoff.

Husserl, Edmund. 1973. In *Zur Phänomenologie der Intersubjektivität. Texte aus dem Nachlass. Erster Teil. 1905–20*, ed. Iso Kern. The Hague: Martinus Nijhoff.

Husserl, Edmund. 1974. In *Formale und transzendentale Logik. Versuch einer Kritik der logischen Vernunft*, ed. Paul Janssen. The Hague: Martinus Nijhoff.

Husserl, Edmund. 1976a. In *Ideen zu einer reinen Phänomenologie und phänomenologischen Philosophie. Erstes Buch: Allgemeine Einführung in die reine Phänomenologie*, ed. Karl Schuhmann. The Hague: Martinus Nijhoff.

Husserl, Edmund. 1976b. In *Die Krisis der europäischen Wissenschaften und die transzendentale Phänomenologie. Eine Einleitung in die phänomenologische Philosophie*, ed. Walter Biemel. The Hague: Martinus Nijhoff.

Husserl, Edmund. 1983. *Ideas pertaining to a pure phenomenology and to a phenomenological philosophy. First book: General introduction to a pure phenomenology*. Trans. F. Kersten. The Hague: Martinus Nijhoff.

Kuhn, Helmut. 1940. The phenomenological concept of the 'horizon'. In *Philosophical essays in memory of Edmund Husserl*, ed. Marvin Farber. Cambridge, MA: Harvard University Press.

Kwan, Tze-Wan. 1990. Husserl's concept of horizon: An attempt at reappraisal. In *Analecta Husserliana*, vol. XXXI, ed. Anna-Teresa Tymieniecka. Dordrecht: Kluwer Academic Publishers.

Landgrebe, Ludwig. 1952. *Philosophie der Gegenwart*. Bonn: Athenaeum Verlag.

Landgrebe, Ludwig. 1963. *Der Weg der Phänomenologie*. Gütersloh: G. Mohn.

Marion, Jean-Luc. 1998. *Reduction and givenness. Investigations of Husserl, Heidegger, and Phenomenology*. Trans. Thomas Carlson. Evanston: Northwestern University Press.

Marion, Jean-Luc. 2002. *Being given. Toward a phenomenology of givenness*. Trans. Jeffrey Kosky. Stanford: Stanford University Press.

Merleau-Ponty, Maurice. 1968. *The visible and the invisible*. Trans. Alphonso Lingis. Evanston: Northwestern University Press.

Merleau-Ponty, Maurice. 1976. *Phenomenology of perception*. Trans. Colin Smith. Atlantic Highlands: The Humanities Press.

Scherner, M. 1974. Horizont. In *Historisches Wörterbuch der Philosophie*, vol. 3, ed. Joachim Ritter. Basel/Stuttgart: Schwabe Verlag.

Walton, Roberto. 1991. Nature and the 'primal horizon'. In *Analecta Husserliana*, vol. XXXIV, ed. Anna-Teresa Tymieniecka. The Hague: Kluwer Academic Publishers.

Walton, Roberto. 1997. World-experience, world-representation, and world as an idea. *Husserl Studies* 14(1): 1–20.

Walton, Roberto. 2002. The phenomenology of horizons. In *The reach of reflection: Issues for phenomenology's second century*, ed. Steven Crowell, Lester Embree, and Samuel J. Julian. West Harford: Electron Press.

Walton, Roberto. 2003. On the manifold senses of the horizonedness. The theories of E. Husserl and A. Gurwitsch. *Husserl Studies* 19: 1–24.

Welton, Donn. 1983. *Origins of meaning*. The Hague: Martinus Nijhoff.

Welton, Donn. 2000. *The other Husserl: The horizons of transcendental phenomenology*. Bloomington: Indiana University Press.

Welton, Donn. 2003. The systematicity of Husserl's transcendental philosophy: From static to genetic method. In *The other Husserl*, ed. Donn Welton. Bloomington: Indiana University Press.

Part I
The Emergence of the Horizon

Chapter 2
Indexicality as a Phenomenological Problem

2.1 The Emergence of Indexicality in Phenomenology and the Immediate Suppression of Its Phenomenological Sense

The task of this chapter is twofold. On the one hand, keeping the above-quoted passage from *Formal and Transcendental Logic* in mind (See Sect. 1.5.1), here I will aim to show how the discovery of the horizon enables phenomenology to rethink what constitutes the problem of the indexicals. On the other hand, and more importantly, the very problem of the indexicals will serve as an entryway into the horizon-problematic in Husserl's phenomenology. I should emphasize from the start that such a procedure will not lead to the realization that the horizon-problematic solves the problem of the indexicals. Taken in isolation from other themes, the concept of the horizon cannot generate such a solution. Rather, the discovery of the horizon forces phenomenology to abandon the manner in which the problem of the indexicals has been formulated in the *Logical Investigations* and to rethink what constitutes the problem's phenomenological character. According to my central thesis that I will present in this chapter, *the phenomenological problem of the indexicals, which the discovery of the horizon gives rise to, is the problem of how any kind of discourse emerges from the pre-predicative experience of the world.* Yet I should emphasize early on that, in the context of this chapter, this claim will only be defended in part. Its full defense presupposes the subsequent analysis in the remaining chapters of this work as a whole.

What are the indexicals? This notion is not to be found in Husserl's writings. Yet if one sets philosophical jargon aside, Husserl turns out to be one of the first philosophers to have addressed the so-called problem of the indexicals under the heading of the "essentially occasional expressions." Especially in the second half of the twentieth century, the problem of the indexicals became an important issue in a number of philosophical circles. By indexicals, or "essentially occasional expressions," we are to understand words like "you," "it," "here," or "then," that is, terms which essentially depend upon some kind of pointing or indicating. Consider, for instance,

S. Geniusas, *The Origins of the Horizon in Husserl's Phenomenology*, Contributions to Phenomenology 67, DOI 10.1007/978-94-007-4644-2_2,

the following phrase: "remember that coffee I used to make while living here? It had a similar flavor to the one he offers you now." Clearly, the meaning of the noun "coffee" is established differently than that of the occasional expressions "that," "he," "here," or "now." In the case of objective expression such as "coffee," the meaning of the term is largely the same when employed by different subjects at different times and places; in the case of the indexical expressions, the meaning depends upon subjective contexts of use. So if the noun "coffee" is uttered by you or by me, in Asia, Europe or North America, at a coffee-shop or in a living room, its meaning remains largely the same; yet the meaning of the whole sentence, due to the presence of occasional expressions, varies accordingly.

Occasional expressions emerge as problematic in Husserl's phenomenology in the context of clarifying the employment of language in *scientific discourse*. This clarification is geared toward the apprehension that language employed in such a context can be objective despite its palpable fusion with the contingent contexts within which scientific claims emerge. The task is that of showing that even though scientific propositions are *context-emergent*, they nonetheless are not *context-dependent*. For such a purpose, the distinction between an *act of meaning-intention*, which is recognized as contingent, and the *meaning of this act*, whose sense is independent from contingency, proves crucial. For example, my *judging* that $2+2=4$ is in countless ways contingent and merely subjective; yet this contingency notwithstanding, the *judgment* itself is ideal and objective. Husserl believes that only if this distinction stands each and every test can one safeguard the objectivity of science by showing that it rests on a firm logical foundation. This foundation must secure the sense of the objective components of discourse as interrelations of *meanings* (rather than meaning-intentions) and show how they can enter into logical relations, which in their own turn would serve as a ground for scientific statements.

One commonly understands the distinction between *meaning-intention* and *meaning* as the distinction between what is *arbitrary* and *subjective* on the one hand, and what is *ideal* and *objective* on the other. Such an understanding of the distinction is not trouble-free. Arguably, none of the perils are as great as the threat posed by the essentially occasional expressions.

The Aristotelian genus/differentia distinction invites one to think of essentially occasional expressions as having a subjective sense and of non-occasional expressions as being objective. After all, in the case of objective expressions, genus and differentia are determined *attributively*, and therefore the meaning of the expression is independent from the context of its use. Yet in the case of occasional expressions, *a determinate differentium is missing*, and therefore the meaning of occasional expressions is dependent upon the context within which such expressions are used. Whether occasional expressions refer to times ("now," "then"), or places ("here," "there"), or persons ("she," "he"), or things ("it, that"), their meanings rest upon the *familiarity with the contexts of their employment*. It therefore seems that the meaning of these expressions must be *subjective* rather than objective. We see thereby why occasional expressions emerge as problematic: (1) they limit the range of logical reason by indicating a domain that lies outside the logical grasp. (2) Moreover, if occasional expressions prove necessary for the application of scientific propositions (which they do, as Husserl himself was willing to acknowledge and which my subsequent analysis will soon make clear), then they introduce doubts into the very plausibility of scientific reason.

Threatened by the outlined hazards, Husserl finds himself compelled to find a solution to this epistemological predicament. One should not overlook how severe his solution turns out to be. Husserl unhesitatingly abolishes the very distinction between subjective and objective expressions. He does so by proclaiming that the content of each occasional expression is an ideal meaning just as much as the content of an objective expression. Occasional expressions are meaningful insofar as they, like ordinary concepts, can be tokened by diverse particulars. Husserl claims that the viability of such a view is already established by the possibility (which admittedly remains always open) of replacing subjective expressions with objective ones. Let me identify this solution as "the replaceability thesis." With this solution, Husserl proclaims the meaning of occasional expressions secured and the problem solved.

Yet as soon as the issue seems to be settled, Husserl voices his own uncertainties in regard to the solution's plausibility, and in these uncertainties lurks the germ of phenomenological counterevidence to the problem's all-too-quick resolution. Husserl remarks that the "replaceability thesis" is *only ideal* and, in fact, *practically unfeasible*. Just try and erase all occasional words from our language; just try and express each subjective experience in definite and objectively fixed terms—all such attempts, claims Husserl, are futile:

> We shall have to concede that such replacement is not only impracticable, for reasons of complexity, but that it cannot in the vast majority of cases, be carried out at all, will, in fact, never be so capable (LI, 321). We are infinitely removed from this ideal .... Strike out the essentially occasional expressions form one's language, try to describe any subjective experience in unambiguous, objectively fixed fashion: such an attempt is always plainly vain. (LI, 322)[1]

Thus the *ideal* solution to the problem of occasional expressions is coupled with its *practical* impossibility. Yet curiously enough, while the emergence of such counterevidence was abrupt and unexpected, its disappearance is even more impetuous and unforeseen. Without providing any reasons, Husserl brushes off the doubts raised by the solution's practical unfeasibility, asserts the puzzle disentangled, and with unbending obstinacy proclaims the boundless range of objective reason (*Schrankenlosigkeit der objektiven Vernunft*). He proclaims: "everything that is, can be known 'in itself.' Its being is a being definite in content, and documented in such and such 'truths in themselves'" (LI, 321).[2]

---

[1] "Freilich müssen wir dabei zugestehen, daß diese Ersetzbarkeit nicht nur aus Gründen des praktischen Bedürfnisses, etwa wegen ihrer Umständlichkeit, unterbleibt, sondern daß sie in weitestem Ausmaße faktisch nicht ausführbar ist und sogar für immer unausführbar bleiben wird." (Hua XIX/1, 95) "Von diesem Ideal sind wir unendlich weit entfernt... Man streiche die wesentlich okkasionellen Worte aus unserer Sprache heraus und versuche irgendein subjektives Erlebnis in eindeutiger und objektiv fester Weise zu beschreiben. Jeder Versuch ist offenbar vergeblich." (Hua XIX/1, 96)

[2] "Alles, was ist, ist 'an sich' erkennbar, und sein Sein ist inhaltlich bestimmtes Sein, das sich dokumentiert in den und den 'Wahrheiten an sich.'" (Hua XIX/1, 95)

Can such a conceptual framework be said to qualify indexicality as a *phenomenological* problem? Yes, insofar as this position is found in a text that belongs to the phenomenological canon. No, insofar as Husserl's solution does not release, but rather suppresses the problem he himself had introduced. Leveling the difference between subjective and objective expressions for the purpose of securing the objectivity of scientific discourse should not be called a phenomenological solution to the problem at hand, unless, however, the plain fact that such a view appears in a text of the founder of phenomenology is a sufficient reason to characterize it as phenomenological. Needless to say, such reasoning is counter-phenomenological.

2.2 The Emergence of the Horizon and the Modification of the Distinction Between Meaning-Intentions and Meaning

Yet the crossroad at which we find ourselves is not that of following Husserl or taking a different path. The fortitude with which Husserl asserts the spelled-out conclusion in the first of the *Logical Investigations* is no greater than the resolve with which he himself rejects it in his revisions to this work. Thirteen years after its first appearance, Husserl characterizes his earlier treatment of the essentially occasional expressions and his defense of the boundless range of reason with an uncompromising verdict as *an act of violence* (*Gewaltsreich*):

> The way in which the occasional expressions (to which, as seen exactly, belong all empirical predications) have been dealt with is an act of violence—the enforced consequence of the inchoate hold of the "truths in themselves" in the *Prolegomena* (Hua XVII, BXIV).[3]

Unfortunately, Husserl does not elucidate the reasons that underlie his unambiguous change in perspective, and what is even worse, neither here, nor in any other of his works, does he clarify his new perspective either. Yet fortunately, he provides us with a few clues that, if we follow them, hold the promise of disclosing his novel conception of indexicality.

Within such a context, the passage from *Formal and Transcendental Logic*, which I quoted in Sect. 1.5.1, proves to be of outstanding significance. Let me reiterate Husserl's claim: Husserl remarks that the reason why his early analysis of occasional expressions was inconclusive stems from the fact that in the framework of the *Logical Investigations*, phenomenology was not yet in possession of the concept of *horizon-intentionality*. However, just as in the revisions of the *Logical Investigations*, so here as well, Husserl does not explain how the emergence of the horizon transforms the problematic of the indexicals. Let us take a detour into the problematic of

[3] "Die Art, wie sie sich mit den okkasionellen Bedeutungen (zu denen doch, genau besehen, diejenigen aller empirischen Prädikationen gehören) abfindet, ist ein Gewaltsreich—die notgedrungene Konsequenz der unvollkommenen Fassung des Wesens der 'Wahrheit an sich' in den *Prolegomena*." (Hua XVII, BXIV)

the horizon so as to see why its discovery is of such outstanding significance for rethinking the problem of indexicality.

Let us recall the preparatory analysis of the horizon in the Introduction. Building my case on how the notion of the horizon is employed in everyday speech, in Sect. 1.1 I singled out three of this term's dimensions of sense. I suggested that (1) the horizon *delimits* each and every phenomenon, that (2) as a limit, the horizon is *versatile* and *unsurpassable*, and that (3) being correlated with our concrete whereabouts, the horizon is *relative* to our concrete situation and in this sense it is irreducible *subjective*. All these dimensions of sense are crucial for the phenomenological appropriation of the horizon as an everyday word. Yet clearly, as my analysis in Sect. 1.3 has shown, as a philosophical concept, the horizon means something more specific. Now we need to take an additional step and, on the basis of 1.3, proceed further to a more elaborate treatment of the horizon as a philosophical concept. Let me turn to two central dimensions of sense that constitute the horizon as a philosophical notion.

Consider the arresting example Merleau-Ponty, with a rather critical eye on Husserl, provides in his *Phenomenology of Perception*:

> When I look at the lamp on my table, I attribute to it not only the qualities visible from where I am, but also those which the chimney, the walls, the table can 'see'.... I can therefore see an object in so far as objects form a system or a world, and in so far as each one treats the others round it as spectators of its hidden aspects.... The completed object is translucent, being shot through from all sides by an infinite number of present scrutinies which intersect in its depths leaving nothing hidden.[4]

This example forcefully reveals what is distinctive about the phenomenological notion of *appearance*. While it belongs to the very sense of objectivity that it can *appear*, an appearance itself is possible *only within a system of appearances*, within which each mode of appearance refers to others. This reference that each appearance draws to the system of appearances is precisely what constitutes the horizonality of the horizon. We thereby obtain *the first sense* of the horizon as a philosophical notion: *the horizon is an implicit system of references* (*Verweisungshorizont*), *which embraces all appearances, due to which an actual appearance is an appearance of a particular objectivity.*

Yet as we take a closer look at what this system of references entails, we notice that it is primarily the presence of *potential* modes of appearances that are given along with the actual ones. To return to Merleau-Ponty's example, the manner in which the lamp on the table "appears" to the chimney or the walls is not merely entailed in my actual perspective; rather, these non-actual "appearances" *determine* the sense of what it means to see this particular object in front of me as a lamp. The implication of potentiality within actuality reveals the second sense that belongs to the horizon as a philosophical notion: the horizon is *the horizon of validity* (*Geltungshorizont*). That is, the implicit references to potential modes of appearance embrace my actual appearance; and only so does my actual appearance become an appearance of a particular objectivity.

[4] See Merleau-Ponty (1976), pp. 68–69.

Yet the horizon does not exclusively belong to appearances; *the intending of these appearances* is just as horizonal. For instance, not only does the sunset I see through my window appear within a horizon; my *seeing* of the sunset also has its own unique horizons. These horizons are of a temporal nature: each present "now" carries with it the horizon of the past and the future. And just as in the case of appearances the horizons proved to be not only systems of references but also systems of validity, so in the case of intending appearances we face an analogical situation: here the co-presence of potential experiences (*Erlebnisse*) co-determines the sense of the actual ones. For example, to hear a sound that obtrudes upon me is already to experience the silence that precedes and follows it. In this regard, my intending of a sound is always already horizonal, due to which it *endures*, and only due to its endurance, can I grasp an appearing phenomenon.

Thus as a philosophical notion, the horizon binds two dimensions of sense on two distinct levels of manifestation. The levels in question are those of *appearances* and *lived-experience*; the dimensions of sense are those of *reference* and *validity*. In Chap. 7, I will address these themes at a greater depth.

At first glance, this excursus into some of the central aspects of horizon-intentionality might sound more like a digression than a clarification of how the novel sense of indexicality could possibly be unveiled. Yet this apparent digression invites one to rethink (although by no means abandon) the distinction between meaning-intention and meaning—a distinction that lies at the heart of Husserl's early analysis of indexicality. As we have seen, the reason why occasional expressions appear problematic stems from the fact that their meaning seems to be permeated with subjective dimensions of sense, which threaten to relativize the objective status of particular meanings. For this reason, Husserl found himself compelled at any cost to show that the relativity in question belongs only to meaning-intentions and not to meanings themselves. As I have shown above, according to the first of the *Logical Investigations*, essentially occasional expressions are meaningful insofar as they can be tokened by diverse particulars. This means that the essentially occasional expressions are meaningful insofar as one can treat them the same way as objective expressions. Thus in a good sense, the meaning of all expressions is established by transcending the subjective contexts, which from the very start seem to plague expressions with arbitrariness. More precisely, according to the view defended in the *Logical Investigations*, the essentially occasional expressions entail subjective dimensions, yet their meaning is not established on the basis of subjective references. However, the fact that the horizon embraces both appearances and lived-experiences brings to light that the subjective dimensions are to be found both at the level of meaning-intentions and meanings. Thus the horizon-problematic brings to light that the meaning of both occasional and non-occasional expressions is subjective through and through in that it is established precisely on the basis of subjective references, even though these references, as my further remarks will show, need not be plagued with arbitrariness.[5]

[5] The central reason for this is, as my analysis in Chaps. 3 and 5 will show, that the horizon is a specifically *transcendental* notion, and thus the subjectivity implicated in the horizon-problematic is a transcendental, and not mundane subjectivity.

What has just been said need not necessarily mean that the subjective dimensions are present in the same way in occasional and non-occasional expressions. Yet the notion of the horizon forces us to recognize that the differences in question are of *secondary* importance in that subjective dimensions are always already *pregiven* before the question of how expressions obtain the status of ideal objectivities can even be posed. Thus the emergence of the horizon signifies that the subjective dimension of sense envelops *all* expressions; the subjective dimension is *irreducible*.

The transformation of the problematic of indexicality that the horizon-intentionality inaugurates places us at the threshold of what I would like to call *phenomenology's unique contribution to the problematic of the indexicals*. Commonly, our philosophical interest in indexicality rests upon a tacit assumption that occasional expressions constitute a different set of problems than non-occasional expressions.[6] The possibility that the analysis of the indexicals might reveal something essential about expressibility *as a whole* is tacitly dismissed from the start, although the grounds for such a dismissal remain unexplained. Phenomenology, particularly in its Husserlian variant, stands out as a tradition, which aims to overcome this form of "unhappy consciousness" by inquiring into what occasional and non-occasional expressions share. Within this philosophical tradition, to thematize occasional expressions is nothing less than to inquire into the origins from which *all* expressions spring.[7] The emergence of the horizon places us at the threshold of such an undertaking. With this in mind, let us proceed further and see how Husserl's discovery of the *noema* provides such an enterprise with a more precise orientation.

---

[6] A detailed engagement in the widespread discussions of the indexicals in other philosophical contexts lies beyond the scope of this investigation. Yet it needs to be acknowledged that these discussions, both in their classical as well as in its more recent and most influential variants, have almost exclusively been motivated by the need to show what distinguishes the indexicals from non-indexical expressions, rather than by the aspiration to reveal what all expressions share. See, for instance, Bertrand Russell's analysis of the "egocentric particulars" (In Russell 1940, Chap. 7), David Kaplan's analysis of the indexicals in his "Demonstratives" (Kaplan 1989, 481–563), John Perry's "Indexicals and Demonstratives" (Perry 1997) as well as Quentin Smith's "The Multiple Uses of Indexicals" (Smith 1989).

[7] It is interesting to note that such a strategy is characteristic not only of Husserl's, but also of Paul Ricoeur's hermeneutical phenomenology. Ricoeur engages in the problematic of the indexicals under the heading of "shifters." His analysis serves a twofold purpose. First, it provides one with a criterion to distinguish between language as discourse and as a linguistic code. As Ricoeur puts it, "the system or code is anonymous to the extent that it is merely virtual. Languages do not speak, people do." (Ricoeur 1976, 12–13) And secondly, the analysis of shifters enables one to provide a non-psychological account of the utterer's meaning, without any mental entity being hypothesized or hypostatized. This is accomplished in virtue of the realization that discourse itself entails a dimension of self-reference, which becomes particularly clear in the case of the indexicals: "the utterance meaning points back towards the utterer's meaning thanks to the self-reference of discourse to itself as an event." (Ricoeur 1976, 13) Ricoeur's analysis of the indexicals in Strawson's works leads to the further realization that brings his analysis very close to Husserl's: "Language is not a world of its own. It is not even a world. But because we are in the world, because we are affected by situations, and because we orient ourselves comprehensively in those situations, we have something to say, we have experience to bring to language." (Ricoeur 1976, 20–21) It is this recognition of the pregivenness of the world as the ultimate source of expressibility that I would like to qualify as phenomenology's contribution to the problematic of indexicality.

2.3 Noematic Intentionality and the Rejection of the Early Analysis of the Indexicals

The foregoing "subjectivization" of expressibility, which the emergence of the horizon has given rise to, forces one to ask whether or not the horizon can safeguard objectivity from the threat of it being reduced to something that is "merely subjective." An inquiry into the problematic of the *noema*, as thematized in *Ideas I*, in relation to the concept of the intentional act, as addressed in the *Logical Investigations*, enables one to examine this question in some detail. I should emphasize from the start that by following such a path, I do not abandon the guiding thought that the horizon offers a way to rethink the very problem of the indexicals. I turn to the problem of the noema because, as we will soon see, *the noema fixes conceptually what has already been revealed phenomenally in Husserl's analysis of horizon-intentionality.*

Such is the case because in *Ideas I*, within the sections that mark the emergence of the noema (§87–§96), the fundamental problem that confronts phenomenology is that of obtaining the means to *fix* and *analyze* objectivity as a phenomenological residuum. That is, the question is *not* that of showing that objectivity does not lose its sense as objectivity after the performance of the *epochē* and the reductions. That objectivity persists as a necessary element of experience has already been determined in earlier sections of *Ideas I*, viz., *in the sections, which uncover the horizonal framework of consciousness*.[8] For this reason, when the noema is introduced in the third part of *Ideas I*, it emerges as an intrinsic dimension of horizon-intentionality.

Husserl had introduced the basic conceptual distinction that underlies the doctrine of the noema as early as the Fifth *Logical Investigation*: "we must distinguish, in relation to the intentional content taken as object of the act, between *the object as it is intended*, and the *object which* is intended" (LI, 578).[9] However, the manner in which this two-sidedness is interpreted in the works from 1900 to1901 and in 1913 is significantly different. In the *Logical Investigations*, everything that phenomenology has to say about the distinction between the object *which is intended* and the object *as it is intended* pertains to the description of the intentional acts.[10] By contrast, *Ideas I* broadens the phenomenological field beyond the real immanence of phenomenological data: in the wake of the reduction, not only intentional acts, but their

[8] I will turn to a direct analysis of these sections in Chap. 4.

[9] "In Beziehung auf den als Gegenstand des Aktes verstandenen intentionalen Inhalt ist folgendes zu unterscheiden: der *Gegenstand, so wie er intendiert* ist, und schlechthin der *Gegenstand, welcher* intendiert ist." (Hua XIX/1, 414)

[10] Thus the distinction in question is immediately coupled with another distinction: "die Unterscheidung zwischen der *Gegenständlichkeit, auf die sich ein Akt voll und ganz genommen richtet*, und den *Gegenständen, auf die sich die verschiedenen Teilakte richten*, welche denselben Akt aufbauen." (Hua XIX/1, 415) Such a strategic move betrays the methodological decision to account for the different manners of the object's givenness by taking into account only the acts of consciousness. Husserl expresses this unambiguously when, after introducing the mentioned distinctions, he proclaims that: "Ihre weiteren Bedeutungen werden uns in den folgenden Untersuchungen erwachsen, in welchen wir einige wichtige Eigentümlichkeiten des phänomenologischen Inhalts der Akte ins Auge fassen und die in ihnen gründenden idealen Einheiten klären wollen." (Hua XIX/1, 416–417)

intentional correlates also, are shown to belong to phenomenological reflection. So as to recognize the significance of this broadening, one needs to address the manner in which Husserl thematizes intentional acts in the Fifth *Logical Investigation*.

Of central significance is the distinction that Husserl draws between act's *matter* and *quality*. Matter is the moment of the act that makes the object count as *this* object and no other; it determines that the act presents *this* object in *just this way*, i.e., in *these* particular articulations and forms (LI VI, §25). While the matter of the act, also called the "interpretive sense," is that element which first gives it reference to an object, the quality of the act determines its character: it determines whether what is given is intentionally presented as wished, asked, posited, etc. Husserl identifies quality and matter as indispensable moments of the act: act-quality is unthinkable without act-matter, just as act-matter is unthinkable without act-quality (LI V, §21). Taken alongside each other, quality and matter comprise the act's *intentional essence*.

Although the intentional essence is indispensable for each and every objectifying act, it does not exhaust the act's phenomenological description. One must also take into account the act's *intuitive fullness*. Each and every presentation, be it perceptual, memorial, or imaginative, alters the object's manner of appearance, depending on the act's intuitive fullness, or vividness. Phenomena are given to us differently depending on the clearness and definitiveness of their manifestation, or on their becoming "pale in color or lost in the midst" (LI V, §21). And even though such intuitive differences are inessential as far as the givenness of objects is concerned (whether I am *vividly* aware of the back side of the birch tree in front of my window is irrelevant in regard to my consciousness of this object as a tree), they nonetheless directly modify the *manner* of the object's givenness.

Having distinguished between the act's matter, quality, and intuitive fullness, let us turn back to Husserl's guiding concern—the two distinct manners of being-conscious of an object. *While the act's matter accounts for our consciousness of the object which is intended, the intentional essence of the act, taken along with the act's intuitive fullness, accounts for our consciousness of the object as it is intended*: such is the answer that one can derive from the *Logical Investigations* to the questions that concerns the distinction which a decade after the publication of this work will give rise to the doctrine of the noema.

It is widely acknowledged that the problematic of the noema is conceivable only on the basis of the *epochē* and the reduction. Such is the case because the reduction announces a broadening of the phenomenological domain without which noematic considerations simply could not emerge. The reduction signifies that the intention directed toward the object and the object itself, *when taken as it lies within the intention*, are not only correlated with each other but also given with equal evidence within the phenomenologically secured framework.[11] Husserl introduces the notion of the noema so as to cover that side of the correlation which remained unthematized in the *Logical Investigations*: the object, *as far as it lies within the intention*, is nothing other than the noema itself. Put in a general (and therefore, ambiguous)

[11] See Bernet et al. (1993), pp. 95–101.

way, the noema is the sense (*Sinn*) without which the objective directedness of the intentional acts would not be possible.

To this day, the question of how this "transcendence within immanence" is to be determined remains a contestable issue. Is the noema, as an immanent entity in consciousness, a part of the occurring thought? Or does it refer to objects beyond thought? Or is it an abstract ideal meaning (*Sinn*) through which the object is given?[12]

A detailed engagement in this set of questions would take me too far afield. This problematic, which at times has almost led a life of its own, has generated two lines of interpretation, generally known as the West Coast (Follesdal, Dreyfus, Smith and McIntyre) and East Coast (Gurwitsch, Sokolowski, Drummond, Cobb-Stevens) positions. According to the first view, the noema is not what consciousness is directed toward but rather, much like the Fregean *Sinn*, it is an abstract entity in virtue of which consciousness is directed toward objectivities. According to the second perspective, the noema is the object-as-it-is-intended, i.e., it is not an abstract entity, but the object itself, yet conceived only in terms of its givenness within the phenomenological framework. The attempt to mediate between these positions is just as old as these positions themselves.[13]

In the present context, my central concern is not that of expressing sympathies and antipathies in regard to any of these three points of view, but rather that of showing how the noema provides a new conceptual framework to thematize occasional expressions. Thus in what follows, in as much as it is possible, I will stay clear of this conflict of interpretations.

In *Ideas I*, precisely in the sections that introduce the problematic of the noema, we find Husserl's critical reflections on his account of intentional acts as presented in the *Logical Investigations*. Here Husserl proclaims that his foregoing analysis of acts is *one-sided* in that it lends itself to both a noetic and a noematic interpretation:

> Critically it may be remarked here that the concepts of the 'intentional' or the 'cognitive' essence which were established in the *Logical Investigations* are indeed correct but are capable of a second interpretation since they can be essentially understood as expressions not only of noetic but also of noematic essences, and that the noematic interpretation, as carried through there one-sidedly in framing the concept of the judgment in pure logic is precisely not the one to be used… (*Ideas I*, 228–229)[14]

One should not overlook that Husserl's critique is not based on the rejection of the foregoing view as unwarranted, but rather on the realization that the earlier

---

[12] See Moran (2000), pp. 155–160.

[13] See Solomon (1977). An earlier version of this paper was presented in 1969 at the American Philosophical Association meeting as a response to Follesdal's "Husserl's Notion of the Noema," which appeared in print the same year.

[14] "Kritisch ist hier zu bemerken, daß die in den 'Logischen Untersuchungen' festgestellten Begriffe des '*intentionalen*' und '*erkenntnismäßigen*' Wesens zwar korrekt, aber noch einer zweiten Deutung fähig sind, sofern sie prinzipiell als Ausdrücke nicht nur noetischer, sondern auch noematischer Wesen verstanden werden können, und daß die noetische Auffassung, wie sie dort einseitig durchgeführt wurde, für die Konzeption des reinlogischen Urteilsbegriffes … gerade nicht die in Betracht kommende ist." (Hua III/1, 217)

analysis needs to be broadened so as to cover not only the noetic, but also the noematic side of consciousness. Only by means of such a broadening can the analysis presented in the *Logical Investigations* do justice to the distinction drawn between the object *that* is intended and the object *as* it is intended. The introduction of the noema[15] is thus meant to countervail the shortcoming with which the *Logical Investigations* have left us.

For our purposes, of central significance is the distinction between three noematic components. The object given just as it is given in consciousness is what Husserl characterizes as the *noematic sense* (let us say, a flock of geese flying over the river, *just as this flock appears*). Yet each noematic sense, due to its *horizonal* framework, co-intends other manners of the object's appearance, the totality of which is covered by the notion of the *full noema* (e.g., all other non-actual appearances of the flock of geese co-intended in each and every appearance). Finally—and here we touch upon yet another element of *horizonal* consciousness—each and every appearance of objectivity is itself intentionally related to the *noematic core* (the flock of geese as the pure "objective sense," which remains identical in perceptual, memorial, or pictorial acts). The latter is a noematic moment despite its *transcendence* in regard to particular noematic senses and to the full noema. In virtue of this transcendence, intentionality shows itself as a title that covers not only the essential feature of the conscious acts, but also of appearances themselves. We thus obtain the means to speak not only of *noetic*, but also of *noematic intentionality*—a notion of central importance for our purposes.

We need not overlook that these three noematic components closely relate to Husserl's earlier account of the act's matter, quality, and intentional essence. First, the *noematic core* is the noematic correlate of the notion of *matter*. That is, just as I can, in virtue of the act's matter, perceive, remember, or phantasize one and the same birch tree, so the perceived, remembered, or phantasized birch tree can be one and the same in virtue of its noematic core. Secondly, the *noematic sense* is the noematic correlate of *intentional essence*, conceived as the unity of the act's matter and quality. Thus perceiving or imagining the birch tree has its noematic correlate in the birch tree as perceived or imagined. Finally, the notion of the *full noema* is the noematic correlate of the manner in which conscious acts intend the *object as it is given* in manifold presentations, i.e., while tacitly co-intending all the other modes of the object's givenness.[16]

For the purposes of this investigation, it is of essential significance that the distinction between the noematic sense, the full noema, and the noematic core reveals how consciousness, despite the subjective elements of sense that permeate the noema, can have an *identical* objectivity as its correlate. In virtue of objectivity's transcendence in regard to the noematic fluctuations of sense, one can characterize

[15] This concept was first used by Husserl in a pencil draft of *Ideas I* in 1912. See Moran (2000, 156).

[16] These parallels between the noetic and the noematic analyses can be pursued further. See Bernet et al. (1993), pp. 95–101.

the object's identity in terms of its *ideality*: the *noematic core* is given to consciousness as an idea which is always correlatively related to the other noematic elements. Having obtained different dimensions of the noema, we are in the position to see that the subjective dimensions of sense, even when they are inseparable from *meaning* (rather than meaning-intentions), cannot corrupt the identical and ideal nature of those objectivities that *these meanings* intend. Thus even if I were to proclaim "look at these birds flying there," the subjective dimensions of sense which permeate this expression would not relativize the status of objectivity—the flock of geese itself. For such to be the case, these subjective dimensions would also have to relativize the noematic core—the flock of geese taken as a purely objective sense. Yet such simply cannot be the case: appearances are relative not only in regard to subjective *Erlebnisse*, but also in regard to the objective sense which they perspectivally intend. Thus the problematic of the noema invites one to draw a distinction between subjective reference and arbitrariness.

One can now see that the reason why Husserl rejects his early solution to the problem of occasional expressions is not exhausted by the need to admit that not all subjective expressions can be replaced by objective ones. The reason is much more radical. The notion of noematic intentionality indicates that the very opposition between subjective and objective expressions, as it was employed in the *Logical Investigations*, had not been thought through to the end. It is highly intriguing that with the notion of noematic intentionality, we face a reversal of the standpoint from within which phenomenology finds itself compelled to address and thematize expressibility. While in the *Logical Investigations*, as we have already seen, phenomenology starts with the assumption that expressions are for the most part objective and then proceeds to the recognition that there is a class of expressions that seems to fall outside this general characteristic, *Ideas I* forces us to recognize that *all* expressions have an irreducibly subjective origin of sense. The discovery of noematic intentionality signals the exposure of a subjective dimension that equally envelops both essentially occasional and non-occasional expressions. Since the noematic core itself is a necessary correlate of the other noematic moments, the subjective dimensions of sense that envelop the latter emerge as necessary conditions in bringing forth the ideality of objectivity. Thus the reversal of the standpoint of the *Logical Investigations* is no longer threatening, because the opposition between noematic sense, the full noema, and the noematic core indicates that even though each and every expression necessarily has a subjective dimension of sense, its mere presence cannot corrupt the ideality of objectivity.

As seen from the perspective of *Ideas I*, just as formal logic is interested only in the noematic core and not in the fuller noemata of judgment (e.g. what makes one judgment evident and the other blind is *meaningless* from the perspective of *formal* logic), so the objective expressions are those expressions which relate only to the noematic core and remain indifferent to the other noematic moments from which their emergence is inseparable. And just as formal logic calls for a transcendental grounding, could the objective expressions not also be said to be in need of an investigation directed at the phenomenological resources from which they spring? The phenomenological framework of *Ideas I* motivates one to ask whether, in direct contrast to the position defended in the *Logical Investigations*, it is not precisely "objective" rather than "subjective" expressions that appear deficient in that "subjective" expressions bring to light the full noematic configurations of sense in regard to which the "objective" expressions remain indifferent.

The foregoing analysis should have made clear that the problematic of the noema reveals that the distinction between subjective and objective expressions cannot be employed as a presupposition in clarifying the ideality of scientific discourse. The source of this distinction must itself be located within subjectivity. Yet what exactly does this pregivenness of subjectivity mean? And more importantly, how does it recast the very problematic of the indexicals?

2.4 The Hidden Dimension of Horizon-Intentionality and the Sense of Indexicality as a Phenomenological Problem

In my attempt to reconsider indexicality as a phenomenological problem, I have shown that the emergence of horizon-intentionality gives rise to far-reaching transformations that pertain to the problem of occasional expressions. Of central significance was the realization that the noema is an element of horizonality and that the different noematic elements reconfigure the relation between subjective and objective expressions. Keeping these transformations in mind, let us turn back yet again to the above-quoted passage from *Formal and Transcendental Logic*, in which Husserl suggests that due to the absence of the notion of the horizon, the *Logical Investigations* could not provide a satisfactory solution to the problem of the essentially occasional expressions (See Sect. 1.5.1).

This claim can be understood in two significantly different ways. On the one hand, it can be taken to mean that horizon-intentionality, as either thematized in *Ideas I* or as further analyzed in Husserl's later writings, generates the new solution to the problem of the indexicals. On the other hand, this claim could also be understood in a more modest way, viz., as a suggestion that the discovery of the horizon leads to the realization that the problem of the indexicals was imprecisely posed in the *Logical Investigations*, and that for this reason, this problem could not find a fully satisfying solution. On the basis of the foregoing analysis, I would like to suggest that Husserl's claim should be understood not in the first, but rather in the second sense. Horizon-intentionality does not solve the problem of the indexicals, as it was formulated in the *Logical Investigations*; rather, the concept of the horizon gives rise to a novel understanding of what constitutes indexicality's phenomenological problematic.

There are a number of reasons that underlie this suggestion. Consider, first, the plain fact that *Ideas I* does not entail any kind of treatment of the indexicals. For this reason alone, Husserl's later reference to the emergence of the horizon in *Ideas I* is of little help when it comes to a direct solution to the problem of essentially occasional expressions. Secondly, as my foregoing analysis has shown, the emergence of the horizon forces one to abandon the rough distinction between subjective and objective expressions that lies at the heart of Husserl's analysis of the essentially occasional expressions in the first of the *Logical Investigations*. Such being the case, the horizon-problematic cannot constitute an answer to the problem of the indexicals as it was posed in the *Logical Investigations*, since the horizon-problematic invites one to reformulate the problem itself.

As my analysis above has shown, (1) all expressions have subjective origins of sense (Sect. 2.2), and yet (2) such a subject-dependence does not relativize all expressibility (Sect. 2.3). In the face the first of these claims, it is no longer possible to view the "replaceability thesis" as a satisfactory solution to the problem of the occasional expressions. Even if it were possible to replace all essentially occasional expressions with objective expressions—a possibility, which Husserl himself considers practically unfeasible—even then all expressions would still retain a subjective reference. As my foregoing analysis of the noema has shown, any *meaningful* notion of objectivity, including the one that plays a central role in scientific discourse, remains correlated with subjectivity. Yet if such indeed is the case, then one might very well wonder whether the horizon-problematic does not reintroduce the problem of psychologism, to which the *Logical Investigations* aimed to provide a compelling response. The second claim highlighted above answers this objection: the problem of psychologism persists only for so long as one considers what is subjective to be interchangeable with what is arbitrary.[17]

Thus Husserl's remark in *Formal and Transcendental Logic* should be understood as a suggestion that the horizon-problematic recasts the very problem of the indexicals. Yet how exactly should one understand the novel conception of the *problem* of the indexicals? With this question, one stumbles into a new difficulty. As will become clearer on the basis of the subsequent analysis, *Ideas I* gives rise only to a provisional, and by no means a full-fledged notion of the horizon. It should therefore come as small surprise that the reconstruction of the novel sense of the problem of the indexicals cannot be derived from this work alone.

Indeed, soon after the publication of *Ideas I*, Husserl realized that his early analysis of horizon-intentionality was by far not conclusive. Its fundamental weakness consisted in the fact that here the horizon was thematized exclusively on the levels of appearances and lived-experience (hence the inseparable bond that ties the horizon to the noema). *The world-horizon*, as a necessary counterpart of these dimensions, remained within this text unexplored. At the moment, I will characterize the world-horizon only in its most general features; Chaps. 10, 11 and 12 will address this highly significant phenomenological theme in detail.

The world as horizon, as the *total-horizon*, entails dimensions of sense that remain hidden from *Ideas I* in that the world-horizon can in no way be equated with mundane horizons that embrace particular objectivities. Rather, the world-horizon accounts for how each of these objectivities is already pregiven within the unity of experience, due to which we always feel "at home" in the world. To describe the

[17] I will turn to the distinction between what is subjective and what is arbitrary in the context of the horizon-problematic in Sect. 3.3.

world in its unique horizonality is thus to thematize its dynamic structure in its pregivenness, to thematize it in a manner that would reveal how the world itself underlies and gives rise to logical categories employed in scientific discourse, which in its own turn transforms the world into the subject matter of its manifold undertakings. Such a thematization of the world reveals how the subjective ambits of sense can obtain an objective dimension. The question of the pregivenness of the horizon is therefore not merely that of realizing how, for instance, the surface of a planet, composed of numerous atomic and chemical particles, derives its sense from a more rudimentary givenness of a landscape; or how the waste of salt water, composed of hydrogen, oxygen, sodium, and about 70 other chemical elements, obtains its sense from a much more elementary appearing of the sea. The question of the pregivenness of the horizon must also account for how our prescientific experience can be of such a nature that it can lend itself to scientific analyses, and thus to scientific language.

Such a deepening of sense that qualifies horizon-intentionality at long last allows me to offer an answer to what I see as the novel problematic of indexicality. The horizon, conceived as the world-horizon, forcefully reveals what *Ideas I* already hints at: *both subjective and objective expressions have the same origins that lie in subjectivity's prepredicative experience of the world.*[18] The fact that the language employed in scientific discourse has not been able to dispel all subjective components should not be conceived as a threat posed to the objectivity of scientific propositions, and phenomenology should certainly not try to cover up this truly remarkable fact by generating theories that suppress the subjective dimension. Rather, this fact should be conceived as a faint echo of the hidden subjective origins from which scientific discourse springs. Indexicality should not be thought of as a peril but rather as a continuously present clue that can motivate phenomenological investigations into the pregivenness of the world. Occasional expressions have such powers.

The soil of the earth may well be ignored when our interests exclusively pertain to why autumn leaves change color, yet it nonetheless remains the source of life from which different types and shapes of plants and trees emerge. Similarly, when our attention is exclusively absorbed by the differences between subjective and objective expressions, these differences should not overshadow a much more basic fact that *all* expressions have subjective origins of sense. The novel problem of the indexicals is thus first and foremost the problem of releasing the forcefulness with which the indexicals reveal the common origins that all expressions share, and as such it is nothing less than the problematic of how scientific discourse emerges from, and within, the life-world.

[18] The emergence of such a position can be traced back to Husserl's revisions of the Sixth Logical Investigation. In these revisions, the relativity that in the original drafts of this work was claimed to pertain only to occasional expressions, is now further broadened so as to absorb all empirical claims. Significantly, Husserl does not merely speak of *okkasionelle Ausdrücke*, but also of *das Okkasionelle* as such – a subtle shift in terminology that announces the recognition of the significance of the surrounding world. Thus already in 1908–1910 Husserl writes: "Jede Bestimmung eines individuellen Objekts dieser Welt setzt voraus eine Beziehung der Bestimmung auf den okkasionellen Bestand." (Hua XX/2, 373)

Admittedly, while the question of how science emerges from the life-world is central to Husserl's phenomenology, to the best of my knowledge, nowhere in his works has Husserl addressed this fundamental question by following the clue of the indexicals. Yet even though it remains unclear how far Husserl has followed the indicated path, I would nonetheless like to suggest that the outlined set of issues constitutes the phenomenological problem of the indexicals with which the reference to the horizon-problematic has left us.

The task of this chapter has been twofold. On the one hand, I aimed to show how the discovery of the horizon enables phenomenology to rethink what constitutes the problem of the indexicals. On the other hand, the very problem of the indexicals served as an archway that leads to the horizon-problematic. Due to such a roundabout procedure, my foregoing analysis calls for a deeper investigation into the notion of the horizon. After all, everything said so far rests upon those dimensions and transformations of sense that make up the vital core of Husserl's notion of the horizon. Yet before engaging a direct analysis of the emergence of the horizon in *Ideas I*, let us first turn to William James' psychology. It will soon become clear that James' *Principles of Psychology* is one of the main resources that underlies the phenomenological problematic of the horizon. In regard to the problem of the indexicals, the inquiry into James' and Husserl' analyses of the horizon will further substantiate my foregoing claim that the "subjectivization" of all expressibility, which the horizon-problematic gives rise to, need not be understood as its relativization. It needs to be seen that the horizon is a specifically *transcendental* notion, and therefore, the subjectivity involved in this problematic is not mundane, but rather transcendental subjectivity. A dialogue between James and Husserl will bring this point to the fore.

References

Bernet, Rudolf, Iso Kern, and Eduard Marbach. 1993. *An introduction to husserlian phenomenology*. Trans. Lester Embree. Evanston: Northwestern University Press.

Husserl, Edmund. 1969. *Formal and transcendental logic*. Trans. Dorion Cairns. The Hague: Nijhoff.

Husserl, Edmund. 1970. *The crisis of European sciences and transcendental phenomenology. An introduction to phenomenological philosophy*. Trans. David Carr. Evanston: Northwestern University Press.

Husserl, Edmund. 1974. In *Formale und transzendentale Logik. Versuch einer Kritik der logischen Vernunft*, ed. Paul Janssen. The Hague: Martinus Nijhoff.

Husserl, Edmund. 1976a. In *Die Krisis der europäischen Wissenschaften und die transzendentale Phänomenologie. Eine Einleitung in die phänomenologische Philosophie*, ed. Walter Biemel. The Hague: Martinus Nijhoff.

Husserl, Edmund. 1976b. In *Ideen zu einer reinen Phänomenologie und phänomenologischen Philosophie. Erstes Buch: Allgemeine Einführung in die reine Phänomenologie*, ed. Karl Schuhmann. The Hague: Martinus Nijhoff.

Husserl, Edmund. 1983. *Ideas pertaining to a pure phenomenology and to a phenomenological philosophy. First book: General introduction to a pure phenomenology.* Trans. F. Kersten. The Hague: Martinus Nijhoff.

Husserl, Edmund. 1984. *Logische Untersuchungen,* Band II/1, ed. Ursula Panzer. The Hague: Martinus Nijhoff.

Husserl, Edmund. 2000. *Logical investigations*, vols. I and II. Trans. J.N. Findlay. New York: Humanity Books.
Husserl, Edmund. 2005. In *Logische Untersuchungen. Ergänzungsband. Zweiter Teil. Texte für die Neufassung der VI. Untersuchung: Zur Phänomenologie der Ausdrucks und der Erkenntnis (1893/93-1921)*, ed. Ursula Panzer. The Hague: Martinus Nijhoff.
Kaplan, David. 1989. Demonstratives. In *Themes from Kaplan*, ed. J. Almog, J. Perry, and H. Wettstein. New York: Oxford University Press.
Merleau-Ponty, Maurice. 1976. *Phenomenology of perception*. Trans. Colin Smith. Atlantic Highlands: The Humanities Press.
Moran, Dermot. 2000. *Introduction to phenomenology*. London/New York: Routledge.
Perry, John. 1997. Indexicals and demonstratives. In *Companion to the philosophy of language*, ed. Robert Hale and Crispin Wright. Oxford: Blackwells Publishers Inc.
Ricoeur, Paul. 1976. *Interpretation theory: Discourse and the surplus of meaning*. Fort Worth: Texas Christian University Press.
Russell, Bertrand. 1940. *An inquiry into meaning and truth*. New York: Norton Company.
Smith, Quentin. 1989. The multiple uses of indexicals. *Synthese* 78: 167–191.
Solomon, Robert. 1977. Husserl's concept of the noema. In *Husserl: Expositions and appraisals*, ed. Frederick Elliston and Peter McCormick. Notre Dame: University of Notre Dame Press.

Chapter 3
James and Husserl: The Horizon as a Psychological and a Philosophical Theme

3.1 William James and the Fringe of Consciousness

How did the horizon become a *philosophical* notion and a *philosophical* theme? It is not possible to answer this question without addressing the relation between Husserl's phenomenology and William James' *Principles of Psychology*. As I have already remarked in Sect. 1.5.1, Husserl himself has repeatedly acknowledged James' *Principles* as the founding text of the horizon-problematic. Consider in this regard Husserl's observation in the *Crisis*: "as much as I know, James was the only one who, under the heading of 'fringes,' became aware of the phenomenon of the horizon—but how could he inquire into it without the phenomenologically acquired understanding of intentional objectivity?" (Hua VI, 267).[1] Moreover, consider this remark in relation to an observation made by Dorion Cairns:

> In 1894 Stumpf called Husserl's attention to James' *Psychology*, and Husserl felt on reading it that James was on the same track as he. The notion of horizon and many others he found there. He had planned to publish a series of articles in the *Philosophische Monatshefte*, but he published only the first, and decided to wait to see what James had done. (Cairns 1976, 36)

Husserl familiarized himself with James in the 1890s and, as Ludwig Landgrebe also remarked.[2] Husserl had already then spoken of James' fringes of consciousness. One thus wonders, why does the discovery of the horizon not take place until 1913? What sense are we to make of a silence that extends for as many as 19 years? These questions make it patently clear that the short path to the origins of the horizon, which begins with *Ideas I*, remains insufficient—and not only because this short path does not take into account the problematic of the "essentially occasional expressions."

---

[1] "James war, soviel ich weiß, der einziger, der unter dem Titel *fringes* auf das Horizontphänomen aufmerksam wurde, aber wie konnte er es ohne das phänomenologisch gewonnene Verständnis der intentionalen Gegenständlichkeit [...] befragen?" (Hua VI, 267)

[2] See Landgrebe (1973).

S. Geniusas, *The Origins of the Horizon in Husserl's Phenomenology*, Contributions to Phenomenology 67, DOI 10.1007/978-94-007-4644-2_3,

These questions bring to light that an inquiry into the origins of the horizon will remain incomplete for as long as one does not address the relation of Husserl's notion of the horizon to James' fringes of consciousness.

James' notion of the fringe of consciousness emerges as a critical response to a common assumption held by the rationalists and the empiricists. *According to both schools of thought, within the inner-world of consciousness, there is no impression, or perception of relations.* On the one hand, the rationalists take this assumption to mean that the *extra mentem* reality of relations does not correspond to anything *inter mentem*, and thus that our awareness of relations must be, indeed *can* be, only known to the pure act of intellect or reason. On the other hand, the empiricists, such as David Hume, go as far as to suggest that just as there are no feelings of relations *within* the mind, so there are no relations *outside* the mind either. For this school of thought, relations are to be understood psychologically, i.e., they find their intelligibility within the associative laws. James' response to both approaches is unequivocal: "both Intellectualists and Sensationalists are wrong. If there be such things as feelings at all, *then so surely as relations between objects exist in rerum natura, so surely, and more surely, do feelings exist to which these relations are known*" (James 1950, 245).

This critique notwithstanding, James did not intend to step beyond the main principles laid out in these schools of thought. Rather, as Aron Gurwitsch has already suggested, the target of James' criticism "is a certain narrowness which had developed in the empiricistic tradition, not the basic principles from which this train of thought had sprung….. James' ultimate end is to rehabilitate empiricism, not to depart from it" (Gurwitsch 1985, 320–321).

This rehabilitation takes the form of the realization that the traditional empiricist conceptions of experience account for only "the smallest part of our minds." The traditional empiricist accounts are illegitimately restrictive due to the above-mentioned exclusion of feelings of relation from the inner-world of consciousness. So as to overcome this shortcoming, James draws a distinction between the *substantive* and the *transitive* parts of consciousness, and it is this distinction that must enable us to identify those feelings to which relations are known. What do these notions mean?

Arguably, language constitutes the model according to which the distinction between the substantive and the transitive parts of consciousness is drawn. One could argue that language is composed of nouns that are (or at least can be) accompanied by images, while the function of verbs is that of joining the nouns to each other. So consciousness also has substantive parts that are accompanied by sensory images and transitive parts that are filled with thoughts of relations. The genuinely revolutionary nature of James' conception of consciousness consists in the realization that just as a noun derives its sense from its relation to verbs that surround it (and thus I am capable of picturing not just, let us say, the birch tree but also the birch-tree-shaking-in-the-wind), so the substantive part of consciousness derives its intelligibility from the transitive parts from which it is inseparable.

Consider a small child learning the alphabet. What exactly happens when the child, in the process of reciting "a, b, c, d, e, f, g," utters the letter "d?" At this very moment, are the other letters outside the child's consciousness? There are good reasons to suggest that they are not. This becomes clear when one turns to a somewhat different scenario. Imagine reading or watching a detective story: a dying man says, "I leave my inheritance to…" and at this very moment, his heart stops beating. Now, the question: is this person conscious only of the words he has uttered, or is he conscious of the full sentence he never had the fortune to complete? The answer to this question is unambiguous. Analogically, as James' own example suggests, if I were on the brink of saying, "the pack of cards is on the table," at the moment when I utter the first few words in this sentence, I would be already aware of where the sentence is leading.

To return to the alphabet example, when a child utters the letter "d," she is aware of the letters that preceded it and the letters that will follow. Yet how? Using James' terminology, one could say that *the child is aware of the letter "d" through the substantive part of consciousness; she is aware of the other letters through the transitive parts of consciousness.* Analogically, no matter what object I might be thematically conscious of, I am also dimly aware of other objects, some of which have just disappeared from my field of vision, while others stand at the threshold of a new experience. This distinction between the transitive and substantive parts of consciousness explains how consciousness continues to be aware of those objects that are no longer present in consciousness. Yet how can objects that are no longer present still be in consciousness? For this to be the case, consciousness itself cannot be restricted to the present. And this means that *consciousness, besides being a consciousness of objects, is always a consciousness of the passing of time*. James' notion of the transitive parts of consciousness ultimately means that temporality is the fundamental structure of conscious life. As Gurwitsch has it, "*what underlies the doctrine of the 'transitive states' is a new conception of consciousness, the definition of consciousness in terms of temporality*" (Gurwitsch 1985, 326).

James' notion of the fringe of consciousness is meant to specify how the temporality of consciousness embraces each and every experience. This distinction between consciousness of objects and consciousness of time allows one to avoid a confusion between two highly important concepts in James' psychology: the *stream* of consciousness and the *fringe* of consciousness. While the metaphor of the *stream* first and foremost designates the manner in which consciousness is aware of itself, the metaphors of *fringe*, *halo*, *suffusion*, *horizon*, and *overtone* serve the purpose of qualifying how objects are given to consciousness. These are not two unrelated themes. The qualification of consciousness as a stream needs to answer the objection that deals with the discreteness and discontinuity of objects:

> Does not every sudden shock, appearance of a new object, or change in a sensation, create a real interruption, sensibly felt as such, which cuts the conscious stream across at the moment at which it appears? Do not such interruptions smite us every hour of our lives, and have we the right, in their presence, still to call our consciousness a continuous stream? (James 1950, 239–240)

James' analysis of the fringe of consciousness is meant to answer this objection. His analysis shows the need to distinguish between *objects*, which are discrete and discontinuous, and the *experience of objects*, which is always marked by continuity. Thunder, for instance, is a discrete object that is distinguishable from other discrete objects that precede and follow its emergence in objective time. Yet as James' elegant example shows, "into the awareness of the thunder itself the awareness of the previous silence creeps and continues; for what we hear when the thunder crashes is not thunder *pure*, but thunder-breaking-upon-silence-and-contrasting-with-it" (James 1950, 240). The memory of silence gone and the expectation of silence to come is the fringe or the halo from which the experience of thunder is inseparable.

This is by far not an unusual experience. A color succeeding another is modified by the contrast; silence sounds delicious after noise; in music, one set of sounds alters the feeling of others; and consciousness itself retains, as James has it, "a kind of soreness" as a condition of present consciousness.[3] Consider also what happens when one is interrupted by someone saying "wait!" or "look!"; or what takes place when one tries to remember a forgotten name; or what happens when one is on the edge of saying something; or what consciousness is conscious of in the face of an experience that it recognizes as familiar. As James perceptively remarks, in all these cases, "the significance, the value, of the image is all in this halo or penumbra that surrounds and escorts it—or rather that is fused into one with it and has become bone of its bone and flesh of its flesh" (James 1950, 255).

One of the metaphors James employs repeatedly to qualify the fringe of consciousness is that of the *horizon*. "When very fresh, our minds carry an immense horizon with them …. And in states of extreme brain-fag the horizon is narrowed almost to the passing word" (James 1950, 256). James' description of the fringe of consciousness in the *Principles* is the psychological source from which emerged the philosophical problematic of the horizon. However, let it be noted early on that, as the following section will show, far from being simply borrowed, the horizon underwent significant transformations and ramifications when it became a specifically philosophical theme.

It is not just a question of remaining faithful to the title of James' magnum opus when it comes to qualifying his analyses of the fringe as psychological. These analyses are psychological due to the manner in which they delimit their central theme. As I have already indicated, this delimitation takes the form of a distinction drawn between objects, in themselves discrete and discontinuous, and the experience of objects that is marked by irreducible continuity. I call James' interpretation psychological because, while it in an unprecedented way broadens our understanding of the inner-world of consciousness, it also methodologically limits itself to the analysis of this inner-world and openly blocks access to the possibility that objects themselves might also be fringed. It is this possibility that is further taken up in Husserl's phenomenology.

This operative distinction between the inner-world of experience and the outer-world of objects is not without its problems. On the one hand, as James himself insists, one needs to draw a distinction between thoughts as subjective facts and

[3] See in this regard James (1950), pp. 234–35.

things of which they are aware.[4] This distinction underlies James' analysis of the fringe of *consciousness*. Yet on the other hand, the fringe of consciousness is nonetheless the fringe that embraces *objects* and not the subjective flow of thinking. This becomes particularly clear when James addresses one of the objections raised by an Irish philosopher Thomas Maguire. In his *Lectures on Philosophy* (1885), Maguire interpreted James' notion of the fringe as some sort of psychic material by which sensations, in themselves separate, are made to cohere together. Maguire rejects such a position with a witty remark: James should see that "uniting sensations by their 'fringes' is more vague than to construct the universe out of oysters by platting their beards" (Maguire, 211).[5] To this James responds in his *Principles* by saying that the fringe is part of the *object cognized*. "Some parts—the transitive parts—of our stream of thought cognize the relations rather than the things; but both the transitive and the substantive parts form one continuous stream, with no discrete 'sensations' in it such as Prof. Maguire supposes, and supposes me suppose, to be there" (James 1950, 258).

Yet arguably, James' response is not fully satisfactory. It remains unclear how the fringe can qualify our *experience* of objects (rather than worldly objects themselves) and at the same time embrace the cognized *object* (rather than the subjective stream). One thus wants to ask: what exactly is the relation between the cognized and the worldly object? And how exactly is one to distinguish between the subjective stream and the cognized object? To these questions, James' *Principles* does not respond.

And it does not respond because of its all-too-close relation to the fundamental principles of British empiricism. The ambiguity we here face is the very same that we find inscribed at the heart of Lockean sensations or Humean perceptions. These notions are ambiguous in that they at the same time stand for the sensing and the sensed, the perceiving and the perceived. Or put somewhat differently, these notions are meant to be only subjective, but they inevitably retain an objective dimension. The same is to be said of James' notions of the fringe, which is simultaneously qualified as the fringe of *consciousness* and as the fringe of the *objects* of consciousness.

It is this ambiguity that allows us to understand the passage from Husserl's late and unfinished *Crisis* to which I already referred. Husserl is full of surprise when he asks: how could James discover the horizon-problematic when he did not have a phenomenological understanding of *intentional objectivity*? We are now in the position to understand this question. First, the absence of intentional objectivity indicates the aforementioned ambiguity that surrounds Jamesian fringes: it remains unclear in which sense the fringes are subjective and in which sense they are objective.

[4] "The confusion is between the thoughts themselves, taken as subjective facts, and the things of which they are aware. It is natural to make this confusion, but easy to avoid it when once put on one's guard. The things are discrete and discontinuous; they do pass before us in a train or chain, making often explosive appearances and rending each other in twain. But their comings and goings and contrasts no more break the flow of the thought that thinks them than they break the time and the space in which they lie." (James 1950, 240)

[5] This reference can be found in James (1950, 258). James' further response to Thomas Maguire's critique follows this reference.

Secondly, and more importantly, this passage also intimates that an adequate understanding of the horizonal structure of consciousness brings about a radical reevaluation of the very notion of objectivity: it belongs to the very sense of objectivity, be it qualified as the objectivity of the inner- or of the outer-world, that it always carries with it its own halo or fringes. These two clarifications allow one to say that, as seen from the Husserlian perspective, what James addresses is the horizon conceived as a *psychological* notion (James deals with the horizon in the sphere of the "inner world of consciousness," not in the domain of the world-consciousness as such). As we are still to find out, a specifically phenomenological analysis of this very problem will bring to light the *transcendental* dimensions of the horizonality of experience. At the moment, a brief hint will have to suffice: the interpretation of the horizon as a transcendental notion derives from a more basic realization that the horizon is a figure of intentionality.

As Gurwitsch puts it, "James may be said to have discovered temporality as the fundamental structure of conscious life" (Gurwitsch 1985, 326). This claim, however, needs to be qualified. The temporality in question is psychological through and through in that it qualifies the inner-world of consciousness, but it does not question, restrict, enrich, or in any way qualify objective time. Within the Jamesian framework, objective time remains unthematized while it is nonetheless assumed to be primary.

James, let us recall, now-and-again speaks of the psychologist's fallacy.[6] This is the fallacy that the psychologist commits when he uncritically projects the fruits of his own labor into the subject matter of his own analysis. On the basis of the foregoing analysis, one could suggest that James himself fell victim to a similar fallacy, which one could call the *physicist's fallacy*. So as to see what this fallacy amounts to, consider James' remarks in the context of his analysis of the selective enterprise of consciousness:

> We may, if we like, by our reasonings unwind things back to that black and jointless continuity of space and moving clouds of swarming atoms which science calls the only real world. But all the while the world *we* feel and live in will be that which our ancestors and we, by slowly cumulative strokes of choice, have extricated out of this, like sculptors, by simply rejecting certain portions of the given stuff. Other sculptors, other statues from the same stone! Other minds, other worlds from the same monotonous and inexpressive chaos! (James 1950, 288–89)

If one were to ask what underlies the assumption of such a primordial chaos, the only response the *Principles* provide lies in James' refusal to commit to any metaphysical view. Yet such a refusal is no less (and arguably even more) problematic than an explicit defense of those metaphysical standpoints from which James aimed to depart. What lies at the heart of this refusal to commit is the uncritical assumption of a privileged scientific discourse, which is assumed to have the means to escape the dominance of fringes and in virtue of this kind of freedom to disclose to us "the only real world" (James 1950, 288). At the heart of the assumption of immunity from any metaphysical commitments lies an uncritical acceptance of a pregiven notion of objectivity.

[6] See James (1950), pp. 278–79.

James' analysis suffers from yet another significant shortcoming. James calls us to acknowledge the fringed nature of experience, yet he does not provide us with the means to compare and evaluate the different ways in which a fringed object can manifest itself to different subjectivities. Yet one might ask: is it not a common experience to come to the realization that the framework of sense into which one had enwrapped a particular object was in fact inappropriate to the object itself? And is it so uncommon to criticize others, or to hear others criticize us, for placing a particular problem within a false framework of understanding? However, if the frameworks of sense are said to be "different worlds" within which "different statues" get to be formed by "different sculptors," then how is one to distinguish between different fringes of sense on the basis of their appropriateness or inappropriateness to the object in question? Needless to say, the qualification of the object itself with the "inexpressive chaos" complicates these matters even further.

One of Husserl's central contributions to the problematic of the horizon lies in his explicit realization that the problematic of the horizon renders such a pregiven notion of objectivity indefensible.

3.2 Horizont, Hof, Hintergrund: Husserl's Discovery of the Horizon

In *Ideas I*, the work in which the specifically philosophical analysis of the horizon originates, Husserl uses the terms *Horizont* (horizon), *Hof* (halo), and *Hintergrund* (background) interchangeably. As he writes in §83, "'horizon' has here the same value as did the terms 'halo' and 'background'" (*Ideas I*, 197).[7] As we saw in the last section, all these terms—halo, background, and horizon—are metaphors that James himself uses interchangeably as approximations of the fringe of consciousness. Should one therefore not conclude that Husserl's alleged "discovery" of the horizon is nothing more than an uncritical appropriation of James' fringes?

Yet such a claim would not be satisfactory, if only because it would remain puzzling why Husserl, besides acknowledging the bond that ties his analysis of the horizon to James' analysis of the fringe, would nonetheless identify *Ideas I* as the text in which the problematic of the horizon originates. We thus seem to find ourselves in a dilemma: on the one hand, if we were to agree with those who identify *Ideas I* as the origin of the horizon, we would remain blind to Husserl's indebtedness to James' fringes. On the other hand, if we emphasized Husserl's indebtedness to James, we would overshadow the momentous character of Husserl's analyses of the horizon.

The only way to escape this dilemma is to acknowledge the specificity of the notion of the horizon while simultaneously admitting its dependence upon Jamesian

[7] "Horizont' gilt hier also soviel wie [...] die Rede von einem 'Hof' und 'Hintergrund'" (Hua III/1, 186, Anm. 1).

fringes. In order to do so, it is not enough to see how closely the notion of the horizon is related to the notions of background and halo; one also needs to find a way to extract a dimension of sense that distinguishes the horizon from the other two terms that are so closely related to it. If this were possible, one could then specify the sense in which *Ideas I* is a groundbreaking work. One could then say that even though the horizon-problematic derives from the problematic of the fringe of consciousness, and even though the problematic of the fringe had played a significant role in Husserl's works prior to the publication of *Ideas I*, the latter work is markedly innovative in that it introduces a distinction between the horizon on the one hand, and halo and background on the other. Hopefully, one would thereby be able to show that *Ideas I* is a groundbreaking text in that it for the first time secures a specifically *philosophical* sense of the horizon-problematic.

In the face of the outlined dilemma, it is hard to overestimate the significance of a passing remark Husserl makes in his brief discussion of the arithmetic horizon in *Ideas I* (Hua III/1, §27). Husserl suggests that for consciousness transposed into the mathematical "world," the natural world *remains in the background, even though it no longer functions as a horizon*. "The world is a background for my act-consciousness, but it is *not a horizon*" (*Ideas I*, 55).[8] Yet how exactly are we to understand this distinction between background and horizon? Husserl himself, unfortunately, does not provide us with an explanation. In the remaining part of this section, I would like to argue that the sense of this distinction derives from the *transcendental* framework of *Ideas I*, which significantly distances this work from Husserl's earlier published writings as well as from from James' *Principles*.

In the last section I spoke extensively of the operative distinction between the inner- and the outer-world that guides over James' analysis of fringes. What underlies this distinction is the assumption that transcendent things can, in principle, be known adequately. If one is willing to hold on to this assumption, what sense is one to make of the undeniable fact that our actual understanding of things is inadequate? Admittedly, it remains possible to claim that we are actually aware only of the inner- and not the outer-world. Husserl, however, unequivocally dismisses such a position. In §43 of *Ideas I*, significantly titled "Clarification of a Principle Mistake," Husserl argues against the conception of God as the subject of absolutely perfect knowledge, who, supposedly, possesses what to us finite beings is denied, viz., who possesses an adequate perception of things in themselves. Such a view, Husserl goes on to argue, is *absurd* (*widersinnig*) in that it rests on the assumption that something transcendent can be given as though it were immanent.[9] According to Husserl, it belongs to the very sense of transcendent things that they can be given to us only through appearances, i.e., given only inadequately.

The abandonment of the assumption that transcendent things lend themselves to adequate cognition is of great significance for it indicates an unprecedented broadening of the Jamesian doctrine of fringes. While in James' *Principles* the fringes

[8] "[Die Welt] ist für mein Aktbewußtsein Hintergrund, aber sie ist *kein Horizont*." (Hua III/1, 60)

[9] For the distinction between the transcendent and immanent givenness, see Hua III/1, §42.

qualify only appearances conceived as subjective phenomena, in Husserl's *Ideas I* the horizons are shown to embrace things themselves. In virtue of such a broadening, the problematic of fringes loses its exclusively *psychological* character and obtains *transcendental* and *constitutive* dimensions.[10]

In §44 of *Ideas I*, Husserl argues that a certain *inadequacy* necessarily belongs to the perception of things, an inadequacy that springs from the fact that things can be given to consciousness only "one-sidedly," only through mere appearances.[11] A particular appearance of the thing itself is given to consciousness in such a way that it entails implicit references to the object's other modes of givenness. It is these implicit references that co-determine the sense of the object in question. Only due to this co-givenness of potential modes of appearances is consciousness aware that there is more to the thing than is manifest in its present appearance, i.e., that the givenness of the thing itself implicates a distinction between the thing and its modes of givenness.

In his last and unfinished *Crisis*, Husserl has endorsed this correlation between the object and its manners of givenness as the fundamental theme of his phenomenology.[12] This late recognition is by no means unprecedented. As Husserl puts it still in *Ideas I*, the functional standpoint is central to phenomenology.[13] The functional standpoint is meant to incorporate all the problems that relate to the constitution of the objectivities of consciousness. And as Husserl further explains, "these functional problems concern the way in which noeses, e.g., with respect to Nature, by animating stuff and combining it into manifold-unitary continua and syntheses bring about consciousness of something such that the Objective unity of the objectivity allows of being harmoniously 'made known,' 'legitimated' and 'rationally' determined" (*Ideas I*, 207).[14]

Once interpreted within the framework of the phenomenological reduction, *the functional standpoint proves to be nothing other than the transcendental standpoint.* By "transcendental" Husserl means the standpoint that subjects lived-experiences (*Erlebnisse*) to a "teleological" interpretation so as to extract their sense-giving dimension, i.e., so as to reveal how consciousness synthesizes lived-experiences and thereby gives rise to meaningful objectivities. It therefore should come as no surprise that in §86 of *Ideas I*, a section dedicated to the analysis of the functional problems,

---

[10] One would be in full right to suggest that Husserl in *Ideas I* turns James' analysis of fringes on its head: Husserl shows that it is not the "inner-," but rather the "outer-world" that is irreducibly horizonal. Yet such a position, as my concluding remarks will suggest, significantly curtail Husserl's early analysis of the horizon: according to Husserl of *Ideas I*, the ideal of knowledge remains "horizonless."

[11] "Ein Ding ist notwendig in bloßen '*Erscheinungsweisen*' gegeben, notwendig ist dabei ein *Kern von* 'wirklich *Dargestelltem*' auffassungsmäßig umgegeben von einem *Horizont uneigentlicher 'Mitgegebenheit*' und mehr oder minder vager *Unbestimmtheit*." (Hua III/1, 91)

[12] See Hua VI, §48.

[13] "Der Gesichtpunkt der Funktion ist der zentrale der Phänomenologie[...]." (Hua III/1, 197)

[14] "Sie [die funktionellen Probleme] betreffen die Art, wie z.B. hinsichtlich der Natur, Noesen, das Stoffliche beseelend und sich zu mannigfaltig-einheitlichen Kontinuen und Synthesen verflechtend, Bewußtsein von Etwas so zustande bringen, daß objektive Einheit der Gegenständlichkeit sich darin einstimmig 'bekunden', 'ausweisen' und 'vernünftig' bestimmen lassen kann." (Hua III/1, 196)

Husserl explicitly acknowledges phenomenology's transcendental nature: "in its purely eidetic attitude 'excluding' every sort of transcendence, on its own peculiar basis of pure consciousness, phenomenology necessarily arrives at this entire complex of *transcendental problems in the specific sense*, and *on that account* deserves the name of *transcendental phenomenology*" (*Ideas I*, 209).[15]

Having recognized that Husserl's analysis of the horizon is driven by transcendental concerns, we are in the position to take a closer look at what it means to qualify consciousness as horizonal. We are now in the position to ask the crucial question: what sense is one to make of the realization that the distinction between the thing itself and its mode of givenness is inscribed within the givenness of phenomena? This inscription indicates that consciousness is conscious of the *limits* that pertain to each and every appearance of transcendent objectivity. Moreover, this inscription indicates that the consciousness of limits is itself possible only because consciousness has always already found a way to transgress them: I know an appearance as an appearance only because it is given to me in the context of other modes of givenness. Consciousness transcends the limits of appearance by way of co-intending the horizons of the object's other modes of givenness. The notion of the horizon thereby proves to be inseparable from the notion of limit. One could even say that the co-presence of the horizon, due to its limiting force, is what makes appearance into an appearance, i.e., into one of the infinitely numerous modes of givenness.

One can thereby see how the Jamesian problematic of the fringes of consciousness, once broadened to embrace not only the inner- but also the outer-world, obtains the original nuance inscribed in the Greek word *ὁρίζειν*, from which the notion of the horizon derives. I have already indicated in the *Introduction* that, as a Greek word, the horizon is a line that marks the extremity of the visual field. It is related to the word "to delimit" and from the outset is conceived in the context that covers every delimitation. The notion of the horizon thereby shows itself inseparable from that of limits, and it is the sense of limits, of boundaries, and thus of inadequacy that is at the center of Husserl's analysis of the distinction between *Horizont, Hintergrund* and *Hof.*

Let us try to make sense of this distinction by turning back to Husserl's mathematical example. What could it mean to say that when I enter into the mathematical world, the natural world remains in the background, but it no longer functions as a horizon? I can think of a particular mathematical equation in a café, surrounded by objects and people I have never seen before; I can also think of a mathematical equation while seated in the armchair in my own living room, where I am surround by familiar objects. These different surroundings form different *backgrounds* within which my thought unfolds. Yet the objects in my surroundings *do not form a horizon*, because even though they surround, *they do not determine mathematical objectivities*. Numbers themselves remain unaffected by the different backgrounds within which they are

[15] "In ihrer rein eidetischen, jederlei Transzendenzen 'ausschaltenden' Einstellung kommt die Phänomenologie auf ihrem eigenen Boden reinen Bewußtseins notwendig zu diesem ganzen Komplex der im *spezifischen Sinne transzendentalen Probleme, und daher* verdient sie den Namen *transzendentaler Phänomenologie*." (Hua III/1, 198)

thought. On the other hand, while thinking of any mathematical number, I have to be dimply aware of all other numbers. Never is it possible to think all the numbers explicitly; numerous numbers always remain non-thematic. Thus they also form a background within which my thought unfolds. And yet, this background is significantly different from that of the natural world. All these inexplicit and non-thematic numbers form the *mathematical world*, which functions not just as a background, but also as a *horizon*. The inexplicit and non-thematic consciousness of numbers forms a horizon because I can only be conscious of a particular number through its relation to other numbers. Here there is no room for arbitrariness: the complete absence of the mathematical horizon would erase the possibility of mathematical thought.

Now we are in the position to see what the distinction between *Horizont*, *Hof*, and *Hintergrund* amounts to. The natural world remains in the background once consciousness transports itself into the mathematical "world," but it no longer functions as a horizon, because *the pregivenness of the natural world does not co-determine the sense of mathematical objectivities*. The notion of background and halo are ambiguous in that they can (although they need not) determine the sense of the objectivity in question. The determination of what makes an object into an object is exactly what makes the horizon into a horizon. *The horizon is necessarily a horizon of the irreducible dimensions of sense*, which means: even though the horizons can be, and in fact are, continuously modified, they cannot be lost. Far from merely transforming the sense of the object, such a loss would simply nullify what makes an object into an object. While a phenomenal being must be given through appearances if it is to be phenomenal, an appearance without references to other appearances is no longer an appearance at all. Thus the loss of thc horizonal structure is inconceivable. The loss of background (*Hintergrund*), or halo (*Hof*), on the other hand, is conceivable: their cancellation results in the modification of objectivity's sense, but not in the cancellation of its being. By losing their background or halo, objectivities still remain objectivities, no matter how radical the alteration of sense this loss brings forth.

Thus *the notion of the horizon stands for what consciousness co-intends in such a way that the sense of what is co-intended is inseparable from what makes the thematic objectivity be an objectivity*. By now, we can finally see the dimension of sense that distinguishes the notion of the horizon from those of background and halo. The latter two notions, as they surfaced before the appearance of *Ideas I*, remained ambiguous in that they did not entail a distinction between those aspects of cogivenness which pertain to objectivity's sense and those aspects which are inseparable from what makes objectivity into objectivity.[16] One can thus say: *the notion of fringes is psychological, while the notion of the horizon is transcendental*.

[16] Such being the case, it is understandable how Husserl in his revisions of his lectures *Grundprobleme der Phänomenologie* from 1910/11 can insert the notion of the horizon even though this notion did not surface in the original draft. We find an identical strategy involved in Husserl's revisions of his *Vorlesungen zur Phänomenologie des inneren Zeitbewußtseins* from 1905. Such a revision does not violate the original drafts, but rather clarifies their sense by bringing earlier phenomenology to the level of its more recent achievements, i.e., by introducing a distinction so as to extract the sense which, even though latently, is already present in the texts which precede the publication of *Ideas I*.

3.3 The Transcendental Dimension of the Horizon

How does Husserl's transcendental analysis of the horizon respond to what I earlier identified as a shortcoming in James' psychological analysis of the fringe? As I have suggested above, James' psychological narrative remains deficient in that it does not provide us with the tools necessary to draw meaningful distinctions between different frameworks of sense within which one and the same objectivity could appear. James' analysis of the fringe faces the danger of relativism and it does not provide us with any clues regarding how we are to overcome it. I would like to suggest that the single most significant advancement of Husserl's analysis of the horizon over James' inquiry into the fringe of consciousness consists in showing how it is possible to affirm the horizonal nature of experience without succumbing to a relativization of experienced objectivity.

As we just saw, in Husserl's phenomenology, the distinction between the horizon on the one hand, and halo and background on the other hand, ultimately amounts to a distinction between a context of sense that is *necessary* for the manifestation of a particular objectivity (*Horizont*) and a context of sense that remains *arbitrary* in regard to the objectivity in question (*Hof, Hintergrund*). This distinction in an important way deepens and modifies James' proclamation, which I have already cited earlier: "other sculptors, other statues from the same stone! Other minds, other worlds from the same monotonous and inexpressive chaos" (James 1950, 289)! If one were to follow up with James' metaphors, one could say that the distinction drawn between *Horizont*, *Hof*, and *Hintergrund* allows us to see that the statues in question are not completely unlike each other; that they are not to be found in different worlds; that the "same stone" is not some "inexpressive chaos" that could nonetheless lend itself to a scientific analysis, but rather a dimension of givenness which itself calls for an appropriate horizon of understanding.

If the term "subjective" is taken to mean that the accomplishments of subjectivity, be they visible or hidden, contribute to the manner of the object's manifestation, then not only background and halo, but the horizon also, is subjective. Yet if the term "subjective" is understood as either something arbitrary, or as something left to the subject's discretion, then the horizon is not subjective (while background and halo are subjective). The horizons of which Husserl speaks in *Ideas I* are first and foremost *objective* horizons of sense, i.e., they do not just bespeak the manner in which any objectivity could be wrapped in an arbitrary context of manifestation, but rather point to those dimensions of sense without which a particular objectivity could no longer be an objectivity. One could thus say that the transcendental framework of Husserl's phenomenology charts the middle course between uncritical objectivism and unsophisticated relativism.[17]

---

[17] To illustrate this point, consider gazing at a birch tree through the window. The birch tree can be given in a variety of different backgrounds. The tree is twenty feet away from the building, but it could just as well be closer to it or further away from it. At this very moment, sunshine falls on its

Thus Husserl's early analysis of the horizon leads to the realization that even though there is a sense in which all horizons are subjective, this does not mean that they are all arbitrary. To the general insight that all horizons of sense are relative to subjectivity, Husserl adds a crucial qualification: *the horizons are relative not in regard to mundane subjectivity, but rather in regard to transcendental subjectivity*. So as to qualify this modification, one could say that even though all fringes of sense are subjective, not all of them are equally appropriate to objectivity. The subject-relativity of the horizons does not compromise their objective force. Even more: in the final analysis, only in virtue of such a subject-relativity of the horizons, can the objective sense of the phenomenon be identified and secured.

One can thus say that Husserl's analysis of the horizon in *Ideas I* marks an unprecedented deepening of the Jamesian doctrine of fringes. Such an acknowledgment, however, should not lead one to overlook that, when viewed from the perspective of Husserl's robust analyses of the horizon that he has unfolded in his later research manuscripts and published works, his early notion of the horizon still remains constrained both thematically and methodically. While James' analysis was limited by the operative distinction between the inner- and the outer-worlds, Husserl's account in *Ideas I*, as will become apparent in Chaps. 4 and 5, is limited by the distinction between the merely phenomenal givenness of the transcendent and the absolute givenness of the immanent. Husserl's subsequent broadening of the horizon-problematic in the so-called genetic phenomenology stems from the realization that the immanent givenness of consciousness is no less horizonal than the transcendent givenness of the world and of things. Arguably, once freed from the distinction between the absolute givenness of consciousness and the phenomenal givenness of the world, the horizon reveals itself as truly universal.

However, the task of this chapter has not been that of providing an exhaustive account of the problematic of the horizon in phenomenology but only that of accounting for how this problematic *emerged* in philosophy. My foregoing analysis leads to the conclusion that the problematic of the horizon stems from, but is not

---

bare branches, but I know very well that in a few hours, this tree will be covered in darkness. The wind has torn off its leaves long ago, yet I know that in a few months, I will see this tree in blossom. What I am describing is the givenness of the tree within a subject-relative background of sense. This kind of background is, of course, arbitrary. The tree can be given in a variety of different backgrounds, and still be a tree.

Yet I also see that this living tree has its roots in the earth. This relation between the tree and the earth also forms its background. And yet, could the tree still be a living tree without any relation to the earth? This relation between the tree and the earth is subject-relative, and yet, it is not arbitrary. Moreover, I know that the tree has a side that I cannot see. Yet I am dimly aware that if I were to walk around the tree, I would see it from a different perspective. Thus the backside of the tree is given at the fringe of my consciousness. This backside of the tree is subject-relative, and yet, its givenness is by far not arbitrary.

Thus the subject-relativity inscribed in the problematic of the fringe of consciousness is ambiguous. On the one hand, it can indicate an arbitrary background of appearance; on the other hand, it can also indicate necessary dimensions of sense, without which a phenomenon in question would no longer be what is.

reducible to the problematic of the fringes of consciousness, as thematized by James in his *Principles of Psychology*. Husserl is to be considered the founder of the horizon-problematic in philosophy, but not because he was the first to have thematized the phenomenon of the horizon. James had already done this with great elegance and at a great depth. The significance of Husserl's analyses consists in having depsychologized this problematic and in having disclosed its transcendental and constitutive dimensions.

In summary, if we don't have reasons to distinguish between such notions as the horizon on the one hand and halo and background on the other, then we simply cannot make sense of Husserl's claim that the notion of the horizon originates in his phenomenology; nor do we have reasons to draw a distinction between the horizon on the one hand, and halo and background on the other hand—a distinction that Husserl draws in *Ideas I*. The only way to make sense of this is to draw a distinction between Husserl's notion of the horizon and James' notion of the fringe of consciousness. According to the interpretation I have offered above, while the notion of the horizon is transcendental, the notion of the fringe of consciousness is psychological. Without drawing such a distinction, we miss what is distinctive about Husserl's notion of the horizon.

Before proceeding to the analysis of *Ideas I*, let me remark in passing that, by extension, the discovery of the transcendental dimensions of the horizonality of experience further substantiates one of my central claims in Chap. 2. According to my argument, the horizon-problematic invites one to recognize that both occasional and non-occasional expressions have subjective origins of sense, yet this recognition does not reintroduce the problem of psychologism. If, as my present analysis suggests, the horizon is a transcendental notion, then to think of subject-relativity and arbitrariness as interchangeable would amount to a category mistake.

References

Cairns, Dorion. 1976. *Conversations with Husserl and Fink*. The Hague: Martinus Nijhoff.

Gurwitsch, Aaron. 1985. In *Marginal consciousness*, ed. L. Embree. Athens: Ohio University Press.

Husserl, Edmund. 1976a. In *Die Krisis der europäischen Wissenschaften und die transzendentale Phänomenologie. Eine Einleitung in die phänomenologische Philosophie*, ed. Walter Biemel. The Hague: Martinus Nijhoff.

Husserl, Edmund. 1976b. In *Ideen zu einer reinen Phänomenologie und phänomenologischen Philosophie. Erstes Buch: Allgemeine Einführung in die reine Phänomenologie*, ed. Karl Schuhmann. The Hague: Martinus Nijhoff.

Husserl, Edmund. 1983. *Ideas pertaining to a pure phenomenology and to a phenomenological philosophy. First book: General introduction to a pure phenomenology*. Trans. F. Kersten. The Hague: Martinus Nijhoff.

James, William. 1950. *Principles of psychology*, vol. I. New York: Dover Publications.

Landgrebe, Ludwig. 1973. The phenomenological concept of experience. *Philosophy and Phenomenological Research* 34(1): 1–13.

Chapter 4
The World-Horizon in *Ideas I*

4.1 A Preliminary Indication of the Horizon in Its All-Determining Sense

Still in Sect. 2.4 I suggested that Husserl's analysis of the horizon remains incomplete for as long as it does not address the problematic of the world-horizon. This suggestion should not be taken to mean that the notion of the world is not to be found in *Ideas I*. It rather means that Husserl's early thematization of the world simultaneously reveals and suppresses its phenomenality; that it uncovers it, while failing effectively to disclose it.

Such is the view I will present in this chapter, and I believe this view comes close to the one Husserl himself defended in the late 1920s. In a short text from 1929 (the same year *Formal and Transcendental Logic* was published), Husserl remarks that the major shortcoming of *Ideas I* had to do with the fact that this text left the notion of the universal horizon undetermined:

> Things and the world are given to me in constant validity, and not only on the basis of a restricted perception, which is already invested with horizons; they are rather given on the basis of a consciousness of validity, whose nature is that of a universal horizon-consciousness. This latter theme also called for a critique, provided that I, as above, put into question the authority (*Recht*) of the world-experience, that experience which provides me with the most original sense and right (*Recht*) for my world-certainty in general... All of this refers in fact to all-embracing and difficult investigations, which have been only late carried through in sufficient concreteness. In the first draft of the *Ideas* they were not yet carried out satisfactorily. (Hua III, 599–600)[1]

---

[1] "Dinge und Welt sind für mich in beständiger Geltung, und nicht bloß aus einer beschränkten einzelndinglichen und schon als das mit Horizonten ausgestatteten Wahrnehmung, sondern aus einem Geltungsbewußtsein der Art eines universalen Horizontbewußtseins. Also auch dieses bedarf einer Kritik, sofern ich wie oben in Fragen eingehe, welcher Art das Recht ist, das Welterfahrung für mich hat, die Erfahrung, aus der ich den ursprünglichsten Sinn und das Recht für meine Weltgewißheit überhaupt gewinne[...]. All das weist in der Tat auf umfassende und schwierige Untersuchungen hin, deren zureichende konkrete Ausführung erst spät gelungen ist. Im ersten Entwurf der *Ideen* war sie noch nicht befriedigend durchgeführt." (Hua III, 400)

S. Geniusas, *The Origins of the Horizon in Husserl's Phenomenology*,
Contributions to Phenomenology 67, DOI 10.1007/978-94-007-4644-2_4,

What sense are we to make of this self-critical commentary in relation to the above-quoted passage from *Formal and Transcendental Logic* (See Sect. 1.5.1), which suggests that *Ideas I* has exposed the notion of horizon-intentionality in its all-determining role? Appearances to the contrary, these passages do not contradict each other, for while the first one speaks of *exposing* the all-determining role of horizon-intentionality, the second one adds an important qualification: this exposition was not carried through to its end. Reading these passages alongside each other invites one to say: if it makes any sense to speak of the "all-determining role of horizon-intentionality" in *Ideas I*, then this role, besides being uncovered, was simultaneously suppressed; it was revealed, yet not effectively resolved.

Can such a view be substantiated on the basis of Husserl's analysis of the world-horizon in *Ideas I*? Before testing such a hypothesis, let me briefly turn back to the second question formulated in Sect. 1.5.1: how is one to understand this "all-determining role of horizon-intentionality?" I would suggest that this role should not be identified with any single determination of the horizon, be it the world-horizon, perceptual-horizons, the arithmetic horizon, etc. Each of the approximations of horizon-intentionality that we find in *Ideas I* contributes to the "all-determining role" of the horizon. With this in mind, let us turn to the world-horizon, as it is thematized in *Ideas I*, and try to decipher what could possibly lead Husserl in 1929 to claim that the most fundamental figure of the horizon remained hidden from his early inquiry into the horizonality of experience.

4.2 The World as the Original Figure of the Horizon in *Ideas I*

The dark conscious horizon of indeterminate actuality (*dunkel bewußte Horizont unbestimmter Wirklichkeit*)[2]—to the best of my knowledge, this phrase, highlighted by Husserl himself, signals the emergence of the notion of the horizon not only in *Ideas I*, but also in Husserl's phenomenology in general. We find it in §27 of *Ideas I*, the section that aims to provide a piece of *pure description* of how the world is given to the *natural attitude*. Although the notion of *pregivenness* is not yet employed either in this section, or in *Ideas I* in general (the term Husserl uses here is rather that of *Vorhandenheit*), it is nonetheless exceptionally well suited to capture the manner in which the world is given to the natural attitude. To say that the world is *pregiven* to natural consciousness is to suggest that the givenness of any objectivity presupposes the prior givenness of the world. Husserl provides a number of qualifications of this type of pregivenness: (1) the world-horizon is pregiven as the

[2] It is interesting to note that even though the notion of the horizon makes its emergence in *Ideas I*, in the original translation of this work into English, the notion of the horizon is almost entirely missing. W. R. Boyce Gibson translates this phrase in the following way: "*a dimly apprehended depth or fringe of indeterminate reality*" (102). §82, for instance, entitled "Fortsetzung. Der dreifache Erlebnishorizont, zugleich als Horizont der Erlebnisreflexion" is translated as "Continuation. The Threefold Limit of Experience, as At Once the Limit of Reflexion Upon Experience."

origin from within which each and every object of consciousness springs; (2) it is pregiven as the *infinite spatio-temporal extension* within which each and every object necessarily finds its place; (3) it incorporates not only inanimate objects, but also animal beings; (4) "it is pregiven" means: I myself am an *insider* within it; (5) it is pregiven as an *endless field* of *possibilities* of consciousness: I can always direct my attention to what it entails; (6) it is pregiven as a *practical* horizon, within which inanimate objects and animal beings have value characteristics.

Such a multifaceted pregivenness of the world-horizon is *unique* in that it is the only horizon not of our own making. There are other horizons consciousness can inhabit, yet only the world is the horizon given as having always already been. As soon as consciousness enters a new horizon, it modifies the world-horizon it has "left." Yet in virtue of such a modification, the world-horizon, although not operative, still abides in the background, and one sooner or later returns to it and inhabits it again.

Let us briefly turn back to the problematic of the *arithmetic horizon*—a theme which Husserl addresses in §28 of *Ideas I* and which I have already addressed in Chap. 3. To take on an arithmetic attitude is to enter into a new horizon and a new "world." In the arithmetic attitude, consciousness recognizes numbers as numbers only because it has already performed the transition into a new space of sense. Yet consciousness is simultaneously aware that the arithmetic world itself is possible only due to its own particular spontaneities. And even if within this attitude the natural world recedes into the background, it nevertheless remains "there" for me, although it is no longer operative as a horizon. "Sie ist für mich kein *Horizont*" (Hua III/1, 60), means: numbers are not given to me as worldly objects; and yet, "the natural world remains 'on hand' … *undisturbed by the new attitude*" (*Ideas I*, 55).[3]

By briefly turning back to the problematic of the arithmetic horizon, we obtain an enriched sense of the pregiveness of the natural world. That the latter is pregiven does not simply mean that it is the ground upon which, and the context within which, objects appear. *Its pregivenness also stamps objects as worldly*. The world specifies a dimension of sense, which is inseparable from all the natural objects given to the natural attitude. Yet what does the mark of worldliness mean? The pregivenness of the world is taken by the natural consciousness to mean the world's *anteriority* in regard to the accomplishments of consciousness. The natural consciousness further assumes this anteriority to mean the world's *preexistence*. The "imprint of worldliness" means that objects given to the natural attitude are immediately apperceived as *existent*.

And yet, to describe the pregivenness of the world-horizon to the natural attitude is not to get hold of its phenomenological sense. Far from aiming to provide a solution, this characterization is designed to formulate a problem. We find the formulation of this problem in Husserl's account of the general thesis of the natural attitude: "'the' world is always there as an actuality" (*Ideas I*, 57).[4] This thesis is problematic,

[3] "Die natürliche Welt bleibt dann 'vorhandene,' ich bin nach wie vor in der natürlichen Einstellung, darin *ungestört durch die neuen Einstellungen*." (Hua III/1, 59–60)

[4] "'Die' Welt ist als Wirklichkeit immer da[…]." (Hua III/1, 61)

because as it stamps the world with the character of an *existing actuality*, this thesis itself is not articulated in any particular act or judgment of consciousness. Precisely because it need not be expressed to be operative, its genuine sense escapes the natural attitude. Such a circumstance might be interpreted as an invitation for phenomenology to carry its analyses over beyond the sphere of the acts of consciousness, yet Husserl hastens to add: even though it is not expressed, the thesis of the natural attitude is a *potential* thesis. Here we are on the brink of a critical step *Ideas I* takes, a step which both allows for the transition from the natural to the phenomenological attitude and which simultaneously blocks access to the phenomenological analysis of the world as horizon. Let us take a closer look to see why such is the case.

4.3 The Suppression of the World-Horizon in *Ideas I*

According to Husserl, the thesis of the natural attitude is a *potential* thesis. This means that "we can now proceed with the potential and inexplicit positing precisely as we can with the explicit judgment-positing" (*Ideas I*, 58).[5] Yet as we proceed in such a manner, we should keep in mind that the purpose of this modification is *only* that of obtaining the possibility of placing the thesis itself within brackets. On the one hand, one has to admit that without treating the thesis of the natural attitude as an explicit thesis, phenomenology is not in the position to clarify the natural attitude. Yet on the other hand, while it addresses the thesis of the natural attitude as though it were an explicit thesis, phenomenology should remain sensitive to the fact that it confronts something that is *not* a thesis, viz., that it addresses something that is not established on the basis of the acts of consciousness. Put otherwise, by turning to the problematic of the world-horizon, phenomenology finds itself in front of those levels of experience, which genetically precede the problematic of the acts of consciousness. However, as Husserl himself will subsequently admit,[6] in *Ideas I* these levels of experience evade thematization.

The fact that the natural thesis is not formulated in an act of judgment indicates the natural attitude's lack of clarity in regard to the very ground on which it stands and thereby provides a motive for the performance of the *epochē*. Being guided by this motive and treating the thesis as though it were a *thesis of an explicit judgment*, one obtains the possibility of putting it out of action. Needless to say, the purpose of "disconnecting," or "bracketing," is not that of turning a thesis into its antithesis but rather that of accounting for its sense and origin. Nonetheless, it is highly curious that, while being motivated by the lack of sense of the world's pregivenness, the performance of the *epochē* and the reduction in *Ideas I* does not lead to a phenomenological description of the world-horizon. As it takes the necessary steps to obtain

[5] "Mit der potentiellen und nicht ausdrücklichen Thesis können wir nun genau so verfahren wie mit der ausdrücklichen Urteilsthesis" (Hua III/1, 62).

[6] See Husserl's "Statische und genetische phänomenologische Methode," in Hua XI, pp. 336–345.

the standpoint from within which the world-horizon could manifest itself in its phenomenological sense, *Ideas I, due to these very methodological procedures*, transforms this sense in such a manner that its manifestation is no longer possible. Such is the case because in *Ideas I*, the treatment of the thesis of the natural attitude *as though it were an explicit thesis* is not limited to justifying and enabling the performance of the *epochē* but is also carried over into the domain within which it is to obtain its own phenomenological description. That is, while at the outset, the thesis of the natural attitude is conceived as only a *potential* thesis, at its outcome it is assumed to be an *actual* thesis. Yet arguably, not being an actual thesis, the "thesis" of the natural attitude resists being fully elucidated in the analysis that is directed to the structure of the acts of consciousness, out of which all theses of consciousness originate. This, I would like to suggest, is the reason why one can simultaneously claim that *Ideas I* both uncovers and suppresses the world-horizon. We are in need of a clearer elucidation.

To treat the thesis of the natural attitude as an explicit thesis of judgment is nothing other than to transform it into a position that is constituted in the *act* of consciousness. Such a transformation is ineluctable, since without an explicit formulation of the world thesis, one could never obtain the evidence that it has indeed been disconnected or bracketed. Without transforming the potential thesis into a thesis intended in an *act* of consciousness, one would not be able to proclaim that the general thesis of the natural attitude consists in the assumption that the world as reality is "always already there" (Hua III/1, 61). The formulation of the thesis rests upon the indicated transformation.

Yet one might very well wonder whether the treatment of the thesis of the natural attitude as an explicit thesis lends itself to *epochē* and the reduction. After all, as Husserl himself repeatedly insists, the thesis of the natural attitude is a *general* thesis. While through the general thesis, one is conscious of the world, through the particular theses, one is conscious of objects in the world. Keeping this distinction in mind, one can object to Husserl's methodological procedure and suggest that the general thesis does not lend itself to any kind of a modification. Consider, for instance, those modifications that are known as modalities of judgment (e.g., doubt, possibility, probability or negation). As they transform the givenness of concrete objectivities, they also leave the pregivenness of the world intact.[7] Should one not therefore say that any kind of modification presupposes an unmodified consciousness and that this unmodified consciousness is nothing other than the consciousness that intends the general thesis of the natural attitude?

Husserl, however, has the resources needed to answer this objection. One should not overlook that the distinction between the general and the particular theses is in place *only within the natural attitude*. What Husserl calls the "radical transforma-

---

[7] "Alle Bezweiflung und Verwerfung von Gegebenheiten der natürlichen Welt ändert nichts an der *Generalthesis der natürlichen Einstellung*. 'Die' Welt is als Wirklichkeit immer da, sie ist höchstens hier oder dort 'anders' als ich vermeinte, das oder jenes ist *aus ihr* unter den Titeln 'Schein,' 'Halluzination,' u. dgl. sozusagen herauszustreichen, aus ihr, die—im Sinne der Generalthesis—immer daseinde Welt ist." (Hua III/1, 61)

tion of the natural thesis" is conceivable only on the basis of the abandonment of this very distinction. Only if "our unmitigated freedom" (Hua III/1, 62) is capable of overcoming this distinction can the world become a phenomenological theme. The unprecedented move of the reduction does not lie in retaining the world thesis in its unmodalized "purity," but rather in opening it to a new modification. Yet what kind of a modification? If one were to subject this thesis to the modalities of doubt or negation, the above-mentioned objection would have to be raised. However, in direct contrast to Descartes' strategy of subjecting the world thesis to the modalities of doubt and negation, Husserl places the world-thesis within the *neutrality* modification.[8]

Yet even such an original modification is not without its problems. I would like to suggest that this transformation, as it is carried out in *Ideas I*, severely constrains the breadth of the horizon-consciousness. While in *Ideas I*, the neutrality modification enables phenomenology to account for the constitution of objects in the acts of consciousness, *the world-horizon is neither an object of consciousness, nor is it constituted in the acts of consciousness*. It is indeed true that after the *epochē*, the world reveals itself as a dimension of sense that "colors" each and every intentional act. Yet the possibility of such a "coloring" remains unexplained. I would suggest that the reason for this lack of explanation is as follows: arguably, *the sense of what is pregiven to each and every act of consciousness cannot be uncovered by an analysis which investigates the structures of these acts*. Yet the investigation undertaken in *Ideas I* is limited to the analysis of the noetic-noematic structures of the *acts* of consciousness and for this reason the text cannot uncover the phenomenological sense of the world-horizon.

Taking what has just been said into account, it can still be surprising, but it is no longer incomprehensible, why in *Ideas I*, after the performance of the *epochē*, the phenomenon of the world evades thematization. It is of course true that in §46–§49 Husserl returns to this theme, yet his all-too-short analysis fails to capture the unique trait of the phenomenon of the world, which I will address in much more detail in Part III. In §46–§49 of *Ideas I*, Husserl's task is that of showing how the world announces itself in the horizonality of beings. However, precisely because Husserl's strategy in these paragraphs is that of accounting for the phenomenality of the world through the structures of the *acts* of consciousness, this analysis inevitably shifts from the analysis of the world-horizon to the inquiry into objects' different manners of givenness. This analysis, therefore, misses what distinguishes the world-horizon from those horizons, which encircle singular objectivities. It should not be overlooked that this is precisely the point Husserl raises in the passage I have already

[8] "*Im Zweifelsversuch*, der sich an eine Thesis, und wie wir voraussetzen, an eine gewisse und durchgehaltene anschließt, vollzieht sich die 'Ausschaltung' in und mit einer Modifikation der Antithesis, nämlich mit der '*Ansetzung*' *des Nichtseins*.... Bei Descartes prävaliert diese so sehr, daß man sagen kann, sein universeller Zweifelsversuch sei eigentlich ein Versuch universeller Negation. Davon sehen wir hier ab.... *Wir greifen nur das Phänomen der* '*Einklammerung*' *oder* '*Ausschaltung*' *heraus*, das offenbar nicht an das Phänomen des Zweifelsversuches gebunden, obschon aus ihm besonders leicht herauszulösen ist[...]." (Hua III/1, 63–64)

quoted in Sect. 4.1, a passage in which Husserl speaks of the notion of the horizon as irreducibly two-sided in that it simultaneously stands for the horizons of objects and the *Totalhorizont*, the world-horizon.

The discussion in §46–§49 culminates in Husserl's notorious thought-experiment of the *annihilation of the world*. Let us follow the argument that leads to this hypothesis in some detail, for arguably, it reveal most clearly how *Ideas I* suppresses the horizonal framework of world-consciousness.

The central purpose of Husserl's analysis in these sections is that of establishing that only consciousness can be given indubitably, and thus absolutely. On the ground of such a conclusion, Husserl will be able to justify his claim that with the *epochē* and the reductions, phenomenology wins a new domain of being which embraces the supposedly lost one (i.e., the natural world) in its totality. So as to establish such a conclusion, Husserl finds it necessary to introduce a distinction between immanent and transcendent perception. The argument is of Cartesian origin and in its preparatory steps it does not add much to what we already find in Descartes. While transcendent perception embraces the givenness of *things*, immanent perception specifies the self-givenness of *consciousness*; while transcendent perception cannot guarantee the existence of its object, immanent perception carries within it such an assurance. "What hovers before one may be a mere figment; the hovering itself, the inventive consciousness, is not itself invented and there belongs to its essence, as to any other lived-experience, the possibility of a perceiving reflection which seizes upon absolute factual being" (*Ideas I*, 101).[9]

Husserl, of course, does not rest content having established such a distinction. With a sudden move, he goes on to argue that transcendent perception, besides embracing the givenness of particular objects, also covers the "world of things" (*Dingwelt*), and for this reason exhausts the manifestation of the world itself. Underscoring every word, he writes: "*over against the positing of the world, which is a 'contingent' positing, there stands then the positing of my pure Ego and Ego-life which is a 'necessary,'* absolutely indubitable *positing*" (*Ideas I*, 102).[10] And even though Husserl qualifies his claim by saying that the existence of the world is dubious only in the *logical* sense, that there are no rational motives which lead one to assert its non-existence, nonetheless, the argument is accepted as sufficient to exclude the manifestation of the world from that domain which will be qualified as phenomenological.

Yet so far the exclusion of the world has not been carried through to its end. For Husserl, *the merely logical possibility of the world's non-being must be transformed into a phenomenological possibility.* It is this transformation that reveals most clearly that the world that is excluded from the phenomenological domain is

[9] "Das Vorschwebende mag ein bloßes Fiktum sein, das Vorschweben selbst, das fingierende Bewußtsein ist nicht selbst fingiertes, und zum seinem Wesen gehört, wie zu jedem Erlebnis, die Möglichkeit wahrnehmender und das absolute Dasein erfassender Reflexion." (Hua III/1, 97)

[10] "*Der Thesis der Welt, die eine 'zufällige' ist, steht also gegenüber die Thesis meines reinen Ich und Ichlebens, die eine 'notwendige'*, schlechthin zweifellose ist." (Hua III/1, 98)

not the world as the horizon, but rather the world conceived as the totality of objects.[11] Already the title of §47, "The Natural World as a Correlate of Consciousness" ("Die natürliche Welt als Bewußtseinskorrelat"), indicates that we are dealing with one particular notion of the world. Yet even though Husserl himself indicates that this notion of the world does not exhaust world-consciousness in its totality ("then the result is the correlate of our factual experience, called '*the actual world,*' as *one special case among a multitude of possible worlds and surrounding worlds*" [*Ideas I*, 106]),[12] he nonetheless insists that all possible worlds are correlates of the modification of the experiencing consciousness. As a correlate of consciousness, the notion of the world that Husserl thematizes in these passages is *the totality of objects* whose manifestation can be accounted for employing the same constitutive structures that underlie the givenness of things. Thus it should come as no surprise that Husserl in §47 calls the transcendence of things the *measure* of all rational claims about transcendence.[13] Yet a claim like this one, like the Procrustean bed, is too short to capture the peculiar transcendence of the world-horizon. The two-sidedness of the horizon cannot help but be overlooked once the standard of things becomes the measure of transcendence.

The world as the totality of objects neither is, nor can be, the world as horizon because *the totality of objects is itself established in virtue of the pregivenness of the horizon.* That the notion of the world, operative within these pages of *Ideas I*, itself rests upon a horizonal framework, Husserl could hardly have made it clearer himself: "any hypothetical formulation in practical life or in empirical science relates to this changing but always co-posited horizon whereby the positing of the world receives its essential sense" (*Ideas I*, 107).[14] Yet *Ideas I* does not bring to light how the horizon-consciousness, which underlies the notion of the world as the totality of objects, itself already entail a more rudimentary manifestation of the world as horizon. And even though the notion of the horizon, in its most original appearance, points in the direction of such a manifestation, the possible annihilation of the world, of which Husserl speaks in §49, shows that *Ideas I* does not follow this path. The world whose annihilation leaves consciousness untouched and thus allows one to qualify its existence as absolute is not the world as horizon, but the world as the correlate of consciousness, conceived as the totality of objects.

---

[11] I will directly address this distinction between the world conceived as the totality of objects and the world-as-horizon in the third part of this work. See also Strasser's "Der Begriff der Welt in der phänomenologischen Philosophie," where this distinction plays a central role in delimiting the problematic of the world-horizon.

[12] "Dann ergibt sich das Korrelat unserer faktischen Erfahrung, genannt '*die wirkliche Welt', als Spezialfall mannigfaltiger möglicher Welten und* Unwelten." (Hua III/1, 100)

[13] "Der echte Begriff der Transzendenz des Dinglichen, der das Maß aller vernünftigen Aussagen über Transzendenz ist, ist doch selbst nirgendwoher zu schöpfen, es sei denn aus dem eigenen Wesensgehalte der Wahrnehmung, bzw. der bestimmt *gearteten* Zusammenhänge, die wir ausweisende Erfahrung nennen." (Hua III/1, 89)

[14] "Jeder hypothetische Ansatz des praktischen Lebens und der Erfahrungswissenschaft bezieht sich auf diesen wandelbaren, aber immer mitgesetzten Horizont, durch den die Thesis der Welt ihren wesentlichen Sinn erhält." (Hua III/1, 102)

Now we are in the position to understand what Husserl means when he speaks in the passage quoted in Sect. 4.1 of having neglected the phenomenon of the world in *Ideas I*. This phenomenon was not accounted for because the analysis undertaken did not distinguish between the different senses of the notion of the horizon (i.e., world-horizon and object-horizon) and therefore shifted its inquiry from the theme of the *Totalhorizont* to that of particular horizons of beings. This transference of sense, although unjustifiable, was at the same time unavoidable, since *Ideas I* limited its analysis to the acts of consciousness. While the analysis of the latter is capable of uncovering the horizons that encircle particular objects, it is incapable of uncovering the world-horizon within which the horizons encircling particular objects are operative.

Let me reiterate the argument offered in this chapter in a succinct way:

1. *Ideas I* treats the world-thesis as though it were an explicit thesis.
2. The treatment of this "thesis" as though it were an explicit thesis enables one to perform the *epochē*.
3. Yet after the performance of the *epochē*, the task of clarifying the world-thesis takes phenomenology in the direction of clarifying the structure of the acts of consciousness.
4. Such a clarification of the acts of consciousness cannot explain the world's pregivenness for arguably, the world's pregiveness is not established on the basis of acts, but rather underlies all acts of consciousness.
5. As we are still to find out, while static phenomenology inquiries into the structures of the acts of consciousness, genetic phenomenology aims to uncover the "hidden history" that underlies the acts of consciousness.

Taking these five claims into account, it becomes understandable why Husserl, in the short passage quoted in the beginning of this chapter, would insist that in *Ideas I*, the world-horizon remains unexplored, even though the world is the original figure of the horizon in this work.

The foregoing analysis thereby leads to the conclusion that *Ideas I* for the first time uncovers, yet simultaneously suppresses the horizon in its all-determining sense; that it reveals the world-phenomenon as a problem, and yet fails to resolve it. The universal horizon was uncovered as a problem in the "piece of pure description" of how the world is pregiven to the natural attitude; it was suppressed by the decision to limit the phenomenological analysis undertaken in this text to the study of the different modes of givenness (rather than pregivenness) and the different structures of the acts of consciousness.

Another way to understand why the world-horizon plays such an irreducibly ambiguous role in *Ideas I* is to say that it is a theme of genetic phenomenology, which already emerges in the context of static methodology. It is, as it were, a genetic theme still dressed in static garb.[15] It is a genetic theme, because the possibility

[15] It will be the task of the second part of this work to disclose what it means to characterize the horizon as a genetic phenomenon (See especially Sects. 6.2 and 6.3).

of its phenomenality rests upon our capacity to uncover the prehistory of conscious acts; it is clothed in static attire, because its thematization is limited to the analysis of the acts of consciousness. We thereby obtain an answer to the third question formulated in Sect. 1.5.1. There I asked, how are we to understand that Husserl in 1929 can look back at a text from 1913 and see within it something that it does not possess? What Husserl sees is the emergence of a theme whose phenomenality transcends the methodological constraints within which it emerges and whose full elucidation will call for a transition from static to genetic phenomenology.

In my foregoing analysis, I have characterized the notion of the horizon as irreducibly ambiguous and, due to this characterization, explained the dual position that the world-horizon occupies in *Ideas I*. What my analysis so far did not take into account are the other qualifications of the horizon, namely, the sense of the co-givenness that accompanies the appearance of any kind of objectivity. To this dimension of horizon-consciousness I now turn.

References

Husserl, Edmund. 1950. In *Ideen zu einer reinen Phänomenologie und phänomenologischen Philosophie. Erstes Buch: Allgemeine Einführung in die reine Phänomenologie*, ed. Walter Biemel. The Hague: Martinus Nijhoff.

Husserl, Edmund. 1962. *Ideas: General introduction to pure phenomenology*. Trans. Boyce Gibson. New York: Collier Macmillan Publishers.

Husserl, Edmund. 1966. In *Analysen zur passiven Synthesis: aus Vorlesungs- und Forschungsmanuskripten, 1918–1926*, ed. Margot Fleischer. The Hague: Martinus Nijhoff.

Husserl, Edmund. 1976. In *Ideen zu einer reinen Phänomenologie und phänomenologischen Philosophie. Erstes Buch: Allgemeine Einführung in die reine Phänomenologie*, ed. Karl Schuhmann. The Hague: Martinus Nijhoff.

Husserl, Edmund. 1983. *Ideas pertaining to a pure phenomenology and to a phenomenological philosophy. First book: General introduction to a pure phenomenology*. Trans. F. Kersten. The Hague: Martinus Nijhoff.

Chapter 5
The Structures of Horizon-Consciousness in *Ideas I*

5.1 Epochē and the Self-Showing of the Horizons

With its central focus on the emergence of the horizon, my foregoing analysis revolved around the problematic of the indexicals (Chap. 2), James' psychology (Chap. 3) and Husserl's early notion of the world-horizon (Chap. 4). Such an emphasis on the beginnings came at a high price: by calling attention to Husserl's discovery of the horizon, I have so far left out of account some of the central aspects of horizon-intentionality. As we will see, in *Ideas I*, horizon-intentionality first and foremost qualifies *object-consciousness*. The present chapter aims to analyze and evaluate this principal aspect of horizon-intentionality.

In Chap. 3 I argued that the problematic of the fringe of consciousness in James' *Principles of Psychology* emerges out of a critique of a common assumption shared both by the rationalists and the empiricists. The differences between them notwithstanding, the main representatives of both schools of thought assert that in the inner-world of consciousness, there is no impression, or perception of relations. The emergence of the horizon in Husserl's phenomenology also brings into question this common thesis. However, in contrast to James, Husserl's task is not so much that of showing that relations are to be found both in the inner- and outer-worlds of consciousness, but rather that the horizon-problematic submerges the very distinction between the inner- and outer-worlds. The *intentional* and *transcendental* nature of the horizon-problematic enables phenomenology to overcome this distinction.

The notion of the horizon remains enigmatic for as long as one does not account for how it can simultaneously entail both subjective and objective dimensions of sense. On the one hand, it is the *object* and not the act of consciousness that comes into view along with its horizons. It is, for instance, the birch tree, and not the seeing of the birch tree, that appears in front of the house, on the other side of the road, in front of the river. To ignore this objective dimension would mean to dispense with the phenomenon of the horizon altogether. Yet on the other hand, these objective horizons

S. Geniusas, *The Origins of the Horizon in Husserl's Phenomenology*, Contributions to Phenomenology 67, DOI 10.1007/978-94-007-4644-2_5,

entail a subjective dimension: it is the birch tree *as it is given to consciousness*, i.e., *as a phenomenon* that *appears* within a horizon. To ask if objects still have their own horizons when they are not given to any imaginable consciousness is to perform a category mistake. To pay no heed to this subjective dimension would also equal dispensing with the notion of the horizon.

Being torn between the subjective and the objective dimensions, one is at a loss as to how to characterize the horizon. Is it an element of consciousness, or a component of things themselves? Whichever option one chooses, one faces resistance from the other side, and not being able to overcome it, one is incapable of choosing either. Experience itself suggests that objects are given along with their horizons; yet the horizons seem to belong neither to objects, nor to the subject. Thus the sense of their irreducible co-presence eludes us.

It is important not to overlook that this sense would continue to elude us for as long as one did not come to terms with the *intentional* framework of consciousness. Not without reason does Husserl speak of the horizon as a figure of *intentionality*. *Only insofar as* consciousness is a consciousness-of does it carries with it its own horizons. More precisely, consciousness is horizonal insofar as it intends objectivities; and these intended objectivities have their own horizons insofar as they are given to consciousness. One could also put this otherwise: if consciousness did not intend objectivities, we would lose the grounds to speak of its horizonal framework. Correlatively, if objects could not be given to consciousness, they would not have their horizons either.

It is thus by far not accidental that Husserl is the founder of the horizon-problematic in philosophy—after all, intentionality is one of his central discoveries. It thereby becomes also comprehensible why, as we have already seen in Chap. 3, Husserl would be so surprised to discover the horizon-problematic in James' psychology, and especially astonished to see that in James, the horizon-problematic, under the heading of the fringe of consciousness, emerges in the absence of intentionality. My analysis of the relation between James' fringes of consciousness and Husserl's notion of the horizon culminated in the realization that the horizon is a distinctively *transcendental* notion, while the fringe of consciousness is psychological. The task of this chapter is to further pursue the transcendental dimension of the horizon.

According to one of the central thesis of *Ideas I*, our understanding of intentionality remains distorted for as long as we do not perform the *epochē* and the reductions. These methods are designed to help us disclose the transcendental dimension of intentional consciousness. Yet how can the transition from the natural to the phenomenological attitude be of any help in clarifying the notion of the horizon when it leads to the establishment of consciousness as a phenomenological residuum? According to a rather standard critique of Husserlian phenomenology, the reduction, far from bringing one back to phenomenality that is still uncontaminated by naturalistic misconceptions, is only "unjustifiably reductive" in that it seems to privilege the subjective dimension at the cost of ignoring the objective. By shifting its focus from the natural world to consciousness, it might seem that phenomenology can inquire merely into the subjective components, thereby from the start restricting the horizon from revealing itself in its full phenomenality.

Yet such an objection sounds plausible only for as long as the specifically phenomenological sense of consciousness eludes us. It should not be ignored that Husserl characterizes consciousness as a "peculiar region of being," and even more significantly, as the "phenomenological world" (Hua III/1, 68), which, as he further observes, necessarily remained unknown and almost unforeseen before the establishment of the phenomenological attitude.[1] Consciousness, as the phenomenological residuum, is essentially different from consciousness as it appears within the natural attitude. This difference consists precisely in the fact that in Husserl's phenomenology, consciousness is not only intentional, but also transcendental. That is, besides enclosing its own acts, consciousness also entails and constitutes the *correlates* of these acts. That is, along with "seeing," it embraces the "seen"; along with "touching," the "touched." To be sure, consciousness does not embrace the correlates of its own acts as something *reel*, i.e., as something immanent; it rather entails them as something *transcendent within immanence*. This somewhat clumsy expression is meant to indicate the transcendental aspect of consciousness: it is through "seeing" that the objective sense of the "seen" is constituted, just as through "touching" the "touched" obtains its own objective sense. Only as intentional and transcendental can consciousness be characterized as a phenomenological *world*, which, as Husserl suggests, far from being a mere aspect of the natural world, embraces the latter in its entirety.[2]

One can therefore say: *the horizons belong neither to the natural world, nor to consciousness as conceived from the natural standpoint. They belong exclusively to intentional and transcendental consciousness as it manifests itself after the performance of the epochē. Rather than being a characteristic of nature and natural things, the horizons are indispensable components of lived-experiences and objects constituted in these experiences* (*Erlebnisse*). Consciousness, as the stream of lived-experiences, can have its own horizons because of its correlative structure, which entails the subjective and the objective dimensions of the horizons.

5.2 Self-Givenness, Originary Givenness, and the Pregivenness of the Horizon

Such a conclusion might appear premature and one might be willing to object that the intentional and transcendental dimensions of lived-experiences only account for the possibility, or the plain openness of experience to attain a horizonal set-up.

[1] Here the word "almost" must not be overlooked: the indubitable givenness of the horizons within the natural attitude is an *enigmatic* givenness whose clarification necessitates one to abandon this standpoint. The horizons thereby themselves provide us with a clue of the phenomenological notion of consciousness, conceived as a "phenomenological world."

[2] Even though, as I have argued in Chap. 4, the correlative nature of consciousness is not *sufficient* to thematize the phenomenon of the world, it is nevertheless *necessary* for such a thematization and *essential* for uncovering the phenomenological sense of the horizon.

Yet the horizons, as one knows still within the natural attitude, are not merely possible; they rather form a constant and irreducible dimension of experience. How are we to understand their permanent presence?

The inseparability of the horizons from the appearance of any imaginable objectivity becomes evident as soon as one lays bare what lies hidden within the notion of appearance. In Husserl's phenomenology, this notion undergoes a similar transformation as the notions of givenness, evidence, or being. Just as givenness need not be absolute, nor evidence adequate, nor being independent, so appearance should not be equated with the appearance of a thematic objectivity, but rather given a chance to reveal itself in its other, even though more opaque, modalities. It is of particular importance that the appearance of an object, as a theme to which consciousness directs its intentional rays, *is not its originary appearance*. Rather, thematic appearances both conceal and presuppose more rudimentary modes of apparition.

Consider noticing the ticking of an old and unattended clock on the wall of a public building. How does this clock, which shows such inaccurate time, emerge as a theme of consciousness? Clearly, a wide variety of other objects could have taken the place of the one that, for one reason or another, has been privileged by consciousness and rendered its momentary or lasting theme. Let us for the moment suspend the question regarding motivation, due to which, under particular circumstances, consciousness privileges certain objects over others, even though this question is of great interest and significance for the horizon-problematic. At the moment, I only want to draw attention to *selectivity* and *discrimination*, which are involved in each and every act of consciousness. To experience an object is to *single it out* from the context of other objects, each of which could have taken the place of the one to which consciousness has directed its interest. But what does this mean? It means that *a whole range of objects has already been pregiven to consciousness before any of them has been singled out as "actual,"* i.e., as a particular theme of consciousness.

As Husserl succinctly puts it, "The seizing-upon is a singling out and seizing; anything perceived has an experiential background" (*Ideas I*, 70).[3] Not only does this mean that a potential field of possible objects of attention is pregiven to consciousness' thematic regard; it also indicates that consciousness is *aware* of its selective enterprise. Namely, it is aware not only of the object it thematizes, but also of the other objects co-given in the background of the thematic objectivity. *These other objects appear, even though they are not singled out.* Or as Husserl puts it, "while we are now adverted to the object in the mode of the 'cogito,' all sorts of objects 'appear;' they are 'intended to' intuitively, they flow together into the intuitive unity of a conscious field of objects."[4] We thus need to distinguish between two

[3] "Das Erfassen ist ein Herausfassen, jedes Wahrgenommene hat einen Erfahrungshintergrund." (Hua III/1, 71)

[4] "[...]während wir jetzt dem reinen Gegenstand in dem modus 'cogito' zugewendet sind, 'erscheinen' doch vielerlei Gegenstände, sie sind anschaulich 'bewußt', fließen zu der anschaulichen Einheit eines bewußten Gegenstandsfeldes zusammen." (Hua III/1, 189)

different dimensions of appearance: an appearance of a *thematic objectivity*, and an appearance of *potential objectivities* that surround it. Motivated by the malleable nature of appearances themselves, one is compelled to distinguish between self-givenness (*Selbstgegebenheit*) and originary givenness (*Originärgegebenheit*)—a distinction Husserl draws in §67 of *Ideas I*. While the thematic givenness of objectivity is to be qualified as its self-givenness, its still potential givenness is to be qualified as its originary givenness.

We are already in the position to see why horizons are indispensable for the givenness of thematic objectivities. Such is the case because each act of consciousness has a halo of background-intuitions ("ein Hof von Hintergrundsanschauungen" [Hua III/1, 71]), which form an inseparable part of the conscious act. Since each actual intention entails a plurality of "co-intentions," Husserl has the grounds to insist that "the stream of experiences can never consist of just actionalities" (*Ideas I*, 72).[5] The givenness of the background justifies the claim that a multiplicity of intentions always underlie a singular thematic act of consciousness; and in virtue of this multiplicity, each and every thematic object cannot help but must appear always wrapped in a horizon of other potential and actual objectivities.

The pregivenness of the horizons, i.e., the pregivenness of a potential field of objects of consciousness, enriches our understanding of how consciousness acquires access to objectivities. Precisely because the latter are already pregiven, even though not as themes of consciousness, they can be transformed into full-fledged themes. In virtue of such a transformation, one is justified in speaking not only of a horizon as a domain of pregivenness, but also of horizons of objectivities, i.e., of horizons which belong to particular thematic objectivities. The horizonal structure of consciousness thereby shows itself as Husserl's word for our *openness to the world*.

However, *Ideas I* provides only a glimpse of what the distinction between these different types of givenness amounts to. While potential objectivities are recognized as originary, the question of how they obtain the status of actual thematic objects remains within this work unexplored. I will return to this question in the third part of this work, in the context of the analysis of the world-horizon (Sect. 10.2) and of Husserl's notion of association (Sect. 12.3).

5.3 The Horizon and the Manifestation of Objectivity

Yet what is the status of all those objectivities that consciousness can thematize due to its own horizonal structure? Should one not say that the objects in question are appearances, as this term is conceived in everyday speech, i.e., that the horizons, far from qualifying our openness to the world, block our access to it, stamping each and every appearance as "merely subjective"? Are the objects given to consciousness through the horizons not phenomena and "merely phenomena," for do the horizons

---

[5] "Der *Erlebnisstrom kann nie aus lauter Aktualitäten bestehen.*" (Hua III/1, 73)

not limit appearances to what is only through subjective reasons motivated, and therefore not genuinely objective? In short, what is it that hides behind the "veil of appearances?"

With these questions in mind, let us recall the distinction between the three noematic components, addressed in Sect. 2.3. This distinction indicates a threefold manner in which the notion of objective sense lends itself to interpretation: it can be conceived as a noematic sense, the full noema, and the noematic core. How do the different senses in question affect the pregivenness of the horizon? One might be willing to insist that while the noematic sense and the full noema bind thematic objectivity to its non-thematic pregivenness, the notion of the noematic core is established in virtue of a breach between the objective sense and its pregiven horizon. Thus the sunset still appears as a sunset when I disassociate it from the non-thematic background, co-given with its actual appearance. Whether I directly see it over the hills, remember having seen it through my bedroom window, or phantasize seeing it over the ocean, the sunset still retains its purely objective sense, which Husserl qualifies as the noematic core.

Should one therefore not say that the establishment of the noematic core rests upon the suspension of the horizons that embrace the object's actual, and thus apparently contingent, manifestation? While the noematic sense embraces the actual appearance along with its *actual horizon*, and while the full noema embraces the actual appearance along with all its *possible horizons*, the noematic core *seems* to refer to the givenness of appearances in the absence of any kind of horizons.

If such indeed were the case, one would have to conclude that the horizons denote only contingent and arbitrary dimensions of sense that envelop the manifestation of concrete objectivities.[6] However, a closer look reveals that the noematic core is horizonal no less than the other noematic moments. It may well be the case that the sunset, as a pure objective sense, transcends the concrete horizons that belong to its actual manifestation, *yet once divorced from any conceivable horizonality, the sunset is no longer a sunset at all.* Any attempt to divorce an appearance from horizonality *per se* can only culminate in the destruction of its objective sense. To put it in the language of *Ideas I*, if the noematic core were non-horizonal, one could no longer qualify it as a noematic moment. It would rather be non- or pre-intentional hyletic data, given before being enlivened by the noetic apprehension of sense. Thus the noematic core does not express the absence of horizons, but rather bespeaks the irreducible presence of those horizons, without which any object whatsoever could no longer be an objectivity.

We thereby see that the pregivenness of the horizons can in no way be qualified as the presence of the merely arbitrary components of sense that embrace appearing objectivities. If it is true that objective sense, once divorced from any imaginable consciousness, loses its objectivity, then it must also be admitted that the pregiven

---

[6] Here we once again touch on the same problem we already addressed in Chap. 3, where we inquired into the dimension of sense that distinguishes Husserl's "horizon" from James' "fringes" and "background."

consciousness is nothing other than *horizon-consciousness*. The question of the structures of the horizon thereby turns out to be inseparable from the very meaning of objectivity.[7]

The foregoing inquiry into different senses of horizonality, as it manifests itself in relation to the three noematic components, significantly enriches our understanding of what it means to qualify the horizon as a necessary and irreducible dimension of appearing objectivities. We now see that the horizons, from which the noematic core is inseparable, are significantly different from those that accompany the noematic sense or the full noema. While in the case of the noematic sense and the full noema, the notion of the horizon embraces all dimensions of pregivenness without discriminating between its contingent and necessary components, in the case of the noematic core, the horizon stands only for the necessary dimensions of co-givenness, i.e., for those dimensions of pregiven sense, without which an objectivity could not be an objectivity.

5.4 The Horizon and the "I Can"

I have early on characterized the horizons as indispensable components of lived-experience and substantiated this claim with a reference to the correlational structure of consciousness. However, in the subsequent analysis I privileged one side of this correlation, namely, that of *experiences*, while leaving the side of *experiencing* unthematized. Yet the horizon-intentionality is irreducibly two-sided; to employ the language of *Ideas I*, the horizon has its *noetic* and *noematic* sides.

The irreducibility of *experiencing* is already entailed in the recognition that each and every mode of appearance entails implicit references to a whole system of appearances. So as to extract this irreducible dimension of the horizon, let us consider what takes place when we hear the sound of a violin—an example Husserl himself employs in §44 of *Ideas I*. Undoubtedly, the sound is given to us only through different manners of givenness. These changing modes of appearance "differ in accordance with whether I approach the violin or go farther away from it, in accordance with whether I am in the concert hall itself or am listening through the

[7] Yet if such indeed is the case, then several questions immediately force themselves upon us. First of all, why is it necessary for consciousness to take on the form of an act? If the origins of objectivity are to be located within the horizonal structures of consciousness, then there must be a sense in which horizons "hold access at bay," forcing consciousness to turn towards the objects which in their original background-appearance are already hinted at, even though they are not manifest as thematic objectivities. Secondly, what forces consciousness to obtain a horizonal set-up? There must be a reason which underlies and motivates a particular kind of emergence of the background-appearance of "proto-objectivities" before consciousness transforms them into actual themes. With these questions, we find ourselves at the threshold of the specifically *genetic* notion of the horizon. Since *Ideas I* does not provide answers to these questions, I will not address them until the second part of this text.

closed doors, etc." (*Ideas I*, 96).[8] And even though "no one mode of appearance can claim to be the one that presents the tone absolutely,"[9] nonetheless, "in the concert hall and at the 'right' spot I hear the tone 'itself' as it 'actually' sounds" (ibid).[10] This example brings out a new dimension of sense that the characterization of the experience of things as *inadequate* envelops. What is given to consciousness through its manners of appearance is given inadequately not only because the actual givenness does not exhaust the intentional objectivity, but also because the actual manner of givenness co-intends a *privileged* manner of appearance in which "the object itself, as it actually is," appears. Consciousness co-intends the manner in which the objects appear under "normal circumstances": how, for instance, the visual objects appear in normal daylight. Listening to a piece of music through the closed doors entails a tacit awareness of how, given the right circumstances, this piece would "really" sound; glancing at a photograph over someone's shoulder and bemoaning that, as it were, "I cannot really see it," entails the awareness that under different conditions the picture would be "genuinely" seen. Clearly, the realization that no givenness of transcendent objects can ever be absolute does not require one to grant each and every manner of appearance equal intuitive force.

But what does this co-intending of non-given modes of appearance mean? A closer look reveals that it is not merely an intending of the object's different modes of appearance. Rather, these manifold appearances are themselves possible only because *consciousness co-intends its own possibilities*, which, if chosen, could render the appearance of an object its "truthful" appearance. To know an object as an object means to be aware of its horizon, which in its own turn entails the awareness of what one can do if one does one of the things one can do. *Let us call this awareness the first form of self-consciousness that the horizon-problematic entails*. To employ the language from Husserl's other works, this is an experience of the "I can." It forms an inseparable dimension of horizon-consciousness, because experiencing necessarily takes place in the horizon of possibilities. Consider Henry Pietersma's elegant account of this aspect of horizon-intentionality:

> A subject's awareness of the context of his actual experience involves awareness of himself as endowed with powers he can activate. What is not actually perceived, for example, is in his awareness something that would be perceived, if he should activate such and such powers of movement. In other words, the awareness of context consists in familiarity with the way in which objects of various kinds "respond" in their appearances to kinds of moves he can in principle initiate. (Pietersma 1973, 98)

Or as Landgrebe once put it, the consciousness of the horizon is the "more or less dark awareness that 'I can continue in this direction and thus gain experiences that

[8] Die Erscheinungsweisen "sind andere, je nachdem ich mich der Geige nähere oder von ihr entferne, je nachdem ich im Konzertsaal selbst bin oder durch die geschlossenen Türen hindurch höre, usw." (Hua III/1, 93)

[9] "Keine Erscheinungsweise hat den Anspruch als die absolut gebende zu gelten" (Hua III/1, 93).

[10] "Im Konzertsaale, an der 'richtigen' Stelle höre ich den Ton 'selbst,' wie er 'wirklich' klingt" (Hua III/1, 93).

will confirm or correct my previous experiences'" (Landgrebe 1973, 10). The "I can" thereby reveals itself as a peculiar form of self-experience; peculiar, because it emerges not in virtue of reflective acts that fall within the rubric of what *Ideas I* calls immanent perception, but rather as a necessary implication of the appearance of transcendent objectivities. The "I can" belongs to the horizon-problematic not because consciousness as a field of lived-experiences itself entails a horizonal framework, but rather because the appearance of transcendent objectivities entail references to our dispositional capacities. The "I can" thereby manifests itself as a dimension inseparable from transcendent experience; it thereby indicates that even the horizon-problematic, which takes place in the confines of transcendent perception, is inseparable from self-manifestation.

5.5 The "I Can" and the Primacy of the Practical

For a more elaborate treatment of the "I can," let us turn to Husserl's analysis of *kinaestheses*, for this analysis constitutes the context within which the most significant description of the "I can" unfolds. Kinaestheses can be defined as the *subjective* movements of the *lived-body* (*Leib*). These movements are subjective because they are *not* experienced from *without* (as physical bodies are), but from *within*. They belong to the lived-body for the very same reason: only when the body is experienced from *within* does it reveal itself as kinaesthetic. Thus Husserl employs the notion of kinaesthesis so as to distinguish the lived-body (*Leib*) from the physical body (*Körper*).

As a synonym for kinaesthesis, Husserl also employs the expression the "egological movement."[11] From a naturalistic perspective which suspends the difference between *Leib* and *Körper*, egological movements can be identified as the movements of our sense organs: they are the *involuntary* movements of my eyes or of my head, of my body-posture (*Körperhaltung*) or of my feeling hands. The philosophical import of these movements consists in the realization that *without kinaestheses, neither the world nor objects could be given to us.*[12] That kinaestheses form an irreducible precondition for the perception of the world is clearly expressed in *Ideas III*: "these free acts, far from being accidental, play a completely necessary role; only through them can the constitution of objects be actually carried out" (Hua V, 121).[13] Let us take a closer look why such is the case.

[11] See Hua IV, pp. 158–160, Hua VI, pp. 108–110.

[12] "Wir nennen diese Bewegungen, die zum Wesen der Wahrnehmung gehören und dazu dienen, den Wahrnehmungsgegenstand möglichst allseitig zur Gegebenheit zu bringen, *Kinästhesen*." (EU, 89)

[13] "Diese freien Akte sind nicht außerwesentliche, sie spielen eine durchaus notwendige Rolle, nur mittels ihrer kann sich die Konstitution von Dinggegenständen überhaupt vollziehen." (Hua V, 121)

Husserl did not invent the term "kinaesthesis" himself, but rather borrowed it from the medical and neuroscientific literature.[14] It is commonly suggested that this notion in the phenomenological framework comprises two Greek words: κίνησις, for which Husserl uses the term *Bewegung*, and αἴσθησις, for which he uses the notion *Empfindung*.[15] I would suggest that such an etymological explanation remains insufficient: besides κίνησις and αἴσθησις, the notion of kinaesthesis also entails the moment of στάσις, for which Husserl has on occasion used the expression *die kinästhetische Ruhe*.

Let us begin with the moment of στάσις. It is to be seen as a noetic correlate of a particular noematic sense, when the latter is abstracted from the full noema.[16] Consider gazing at a birch tree in front of the window during an experience of daydreaming. The appearance of the tree is what it is; it does not excite any kind of perceptual interest; it is "just there," appearing in an unchanging mode of givenness. The noetic correlate of such a noematic sense is the kinaesthetic moment of στάσις. As a *kinästhetische Ruhe*, it is the *intending* of the given noematic sense, when the latter neither excites perceptual interest, nor changes its modes of givenness. As a mere intending of what is "just there," the *kinästhetische Ruhe* is an expression of subjectivity's rudimentary and involuntary spontaneity; it is the most basic expression of an "I do," taken in isolation from the "I can."[17] The lived-body that such kinaestheses bespeak is nothing other, and nothing more, than the zero-point of orientation.

Now consider being woken up from daydreaming by noticing that the tree's branches are being shaken by the wind, that its leaves, appearing in a sundry of colors, are being torn away by the unexpectedly obtruding storm. What was initially given as being "just there" suddenly excites perceptual interest: the movement of my eyes and of my body (*Leib*) follows the direction of the shaking branches and the floating leaves. What is given is no longer an abstract noematic sense but a mul-

[14] "The neologism kinaesthesis … was first used by Charlton Bastian who was a doctor and early neuroscientist. With the word 'kinaesthesis' he attempted to explain what he called a 'muscular sense.' Kinaesthesis is, as Bastian pointed out, another expression for the muscular sense which makes us able to lift for example our arm. At the same time when Husserl reintroduced this term, another author, the German doctor Paul Schilder, uses the term kinaestheses." (Brenner 2006, 37)

[15] See, for instance, Ulrich Claesges (1964, 64), or Henning Röhr's account of *Kinästhesis* in the *Wörterbuch der phänomenologischen Begriffe*, 308.

[16] For the distinction between the different noematic moments, see Sect. 2.3.

[17] Consider in this regard Cairns's conversation with Husserl (6/5/32): "I then asked whether the distinction noesis – hyle was a real dichotomy of the immanent constituents of *Erlebnisse*, or whether perhaps kinaesthesis formed a third class. He replied, first of all, that although he had tried to distinguish kinaesthesis sharply from hyle, yet there are in certain cases hyletic concomitants which necessarily accompany the kinaesthesia. I said, suppose we leave these out of account. He said, then all one can say is that the kinaesthesia is the original form of the 'I do.'" (Cairns 1976, 73)

tiplicity of noematic senses. The kinaesthetic *Bewegung* is the noetic correlate of this noematic multiplicity. Such a kinaesthetic consciousness is not merely the consciousness of an "I do," but also of an "I can." The lived-body entailed in such kinaestheses is not merely the zero-point of orientation, but also a body-in-motion.

In both the "static" and the "kinetic" kinaestheses, we can isolate a moment of *αἴσθησις* (*Empfindung*).[18] And precisely because the latter is present in such essentially different types of kinaestheses, *αἴσθησις* is an ambiguous kinaesthetic moment. Ulrich Claesges has noted this ambiguity as many as four decades ago in his *Edmund Husserl's Theorie der Raumkonstitution*.[19] Building on the basis of Claesges' analyses, let us indicate this equivocation by saying that the "aesthetic" moment can be understood in the common sense as "data-sensation" (*Datenempfindung*) and in a less common sense as "sensation of position" (*Stellungsempfindung*). On the basis of the preceding analysis one could say that, on the one hand, what Claesges calls "data-sensation" is nothing other than the moment of *αἴσθησις*, as it manifests itself in "static" kinaestheses: it is the noetic correlate of an abstract noematic sense.[20] On the other hand, one could also say that "sensation of position" is the moment of *αἴσθησις* in "kinetic" kinaesthesis: it is the noetic correlate of a multiplicity of noematic senses. While "data-sensation" belongs to the mere consciousness of an "I do," "sensation of position" entails a consciousness of an "I can." And just as the "I can" belongs to kinaestheses as a *system of possibilities*, so the multiplicity of noematic senses form a synthetic unity in virtue of which consciousness can be in possession of a full noema.

---

[18] The notion of *Empfindung* is exceptionally problematic in Husserl's phenomenology. The general distinction between noesis and noema still leaves out of account a third dimension of consciousness, which Husserl calls *hyletic*. This dimension embraces what is designated by the notion of sensations. The notion of hyle has been rejected by Sartre in his *Transcendence of the Ego*. Sartre argues that this notion has no theoretical validity: It is out of place within phenomenology because it renders inexplicable how consciousness transcends itself to objects in the world. Yet this notion can also be rejected because of its lack of phenomenological validity: The notion of hyle is phenomenologically unjustifiable because it cannot be fulfilled in intuition. We find such a view expressed by Quentin Smith in his "A Phenomenological Investigation of Husserl's Theory of Hyletic Data." According to Q. Smith, hyletic data are not pre-intentional and pre-objective, of independent and real existence, but rather a projection of reflecting consciousness. Dorion Cairns also persistently rejects the problematic of *hyle* in his *Conversations with Husserl and Fink*, primarily basing his position on the lack of evidence in Husserl's works for drawing a distinction between *hyle* and *noesis*.

In the present interpretation, I will leave the highly contestable distinction between the hyletic and noetic dimensions of consciousness out of account. Sensations of which I will speak will always already be noetically "enlivened" sensations.

[19] Ulrich Claesges (1964), pp. 64–66.

[20] "Soll nun das Phantom, wie es hier geschieht, als noematisches Korrelat der Kinästhese qua Noesis aufgesfaßt werden, so können wir das eigentliche Empfindungsdatum gemäß der ersten Gegebenheit der noematischen Sphäre, dem Aspekt, '*Aspektdatum*' nennen." (Claesges 1964, 64)

Yet the expression "multiplicity of noematic senses" is itself ambiguous: it can be understood in terms of the object's inner- and outer-horizon.[21] Not surprisingly, therefore, the expression "sensation of position" can itself be understood in a two-fold way. This ambiguity rests upon a different manner in which the "I can" *as a system of possibilities* lends itself to interpretation. On the one hand, the system in question can be understood in an *objectively determinate way*. It then refers to all the kinaesthetic possibilities that consciousness co-intends in regard to the noematic possibilities of the object's givenness. In this sense, the "I can" is the correlate of the object's inner-horizon. On the other hand, the system of possibilities can be also understood in an *objectively indeterminate way*. It then pertains to the kinaesthetic possibilities that consciousness co-intends in regard to *any* object's givenness. Taken in this objectively indeterminate way, the "I can," as a system of possibilities, is the noetic correlate of the object's outer-horizon.

---

[21] It is somewhat paradoxical that in *Ideas I*, an explicit distinction between the object's inner- and outer-horizons is not yet drawn. We encounter this terminology in Husserl's later works (such as *First Philosophy II*)—works that are less concerned with the horizons of objectivities and more interested in the genetic dimensions of the horizon-problematic. I will address the genetic dimensions of the horizon-problematic in Part II. In the present context I only want to emphasize that from a thematic point of view, the distinction between the object's inner- and outer-horizon has less to do with the genetic dimensions of the horizonality of experience than with a static inquiry into the horizons of objectivities, which is so forcefully presented in *Ideas I*. What is more, even though the distinction between the object's inner- and outer-horizons is not explicitly drawn in *Ideas I*, it is nonetheless operative in this work to such a degree that our understanding of the horizon-problematic in *Ideas I* remains deficient for as long as this distinction is not taken into consideration.

The notion of the inner-horizon is meant to cover those dimensions of sense that belong to the thematically given objectivity, even though these dimensions are not explicitly thematized in consciousness. For instance, the piece of paper that lies on my table has a side unseen, and I must be conscious of this side if I am to identify this object as a piece of paper. Using Husserl's terminology, one would say that the unseen side of this (or any) object is given to me as the object's inner-horizon.

Spatial objects provide a good illustration of Husserl's notion of the inner-horizon, yet one should not be left with the impression that spatiality exhausts the sense of the object's inner-horizons. Temporal objects also have an inner-horizon: the melody that I hear is given to me not only in the now, but also in the past and the future that I must now intend. This implicit givenness of the melody in the past and future qualifies the melody's inner-horizon.

The notion of the outer-horizon is meant to qualify what is co-given to consciousness in such a way that the co-given dimensions of sense constitute other objectivities besides the one, to which consciousness has directed its thematic regard. For instance, the piece of paper, conceived as a theme of consciousness, is *on the desk*, the desk is *in the room*, the room—*in the building*, etc. These co-given backgrounds of sense exemplify Husserl's notion of the object's outer-horizon.

Just as in the case of the inner-horizon, so in the case of the outer-horizon as well, not only spatial, but temporal objects also have their own outer-horizons. To return to the example of the melody, this temporal object itself emerges into the field of consciousness in the temporal horizon of a number of other objects that consciousness has and will thematize.

One more remark on the object's inner- and outer-horizons: *for any object to be given to consciousness, it must have an inner- and outer-horizons*. Put otherwise, what we face here is not a description of some peculiarities of experience, but rather a description of the essential structures of consciousness.

The correlation between the "I can" on the one hand and the object's inner- and outer-horizon on the other indicates the *primacy of the practical* in Husserl's phenomenology. The noematic horizons of sense are thus possible only as far as they are correlated to noetic *possibilities*. The inseparability of the noematic horizons from the "I can" shows that the problematic of the horizon in Husserl's phenomenology is not that of spatial configurations (as the mere indication of the inner- and outer-horizons might suggest), but rather that of the "facultative possibilities" (*Vermöglichkeiten*) of subjectivity. We thereby see why Husserl could be so astonished while hearing how the theme of practical possibilities could have been overlooked and misconstrued in his phenomenology:

> I told him [Husserl] about Becker's recent lectures, wherein he contrasted what he stated to be Husserl's notion of possibility, as "pure possibility" alone, with Heidegger's notion of possibility as potentiality, *Vermögen*. Naturally Husserl was astonished that anyone could attempt a distinction between him and Heidegger along this line. For 15 years at least, he had been operating with the notion of *Möglichkeit als Vermögen*—he had even been using the term *Vermöglichkeit* to express the ego's free potentiality. (Cairns 1976, 4)

5.6 The Horizons of Experience and the Horizon of the Stream of Experience

So far my analysis of the horizon-consciousness was limited to the exploration of the structures that pertain to transcendent perception. Husserl, as is well known, draws an inflexible distinction between transcendent and immanent perception (Hua III/1, §42–§44). In the midst of elaborating the differences between them, he remarks: "everything which we have worked out about the givenness of the physical thing loses its sense here" (*Ideas I*, 96).[22] Such is the case primarily because in the framework of immanent perception the distinction between the object and its manners of appearance is out of place: "an experience is not adumbrated" (*Ideas I*, 90)[23]; "no experience … is presented" (*Ideas I*, 95).[24] Yet the distinction between the object and its manners of givenness was vital in my analysis of horizon-consciousness. Given that Husserl in §44 employs the notion of the horizon only in characterizing transcendent and not immanent perception, one might be inclined to assume that the horizon itself is a peculiar feature of transcendent perception, or correlatively, that the peculiar nature of immanent perception consists in its lack of horizonal structures. Yet a closer look reveals that such is not the case in the least.

Already §44 provides a clue that lived-experiences themselves might entail their own unique horizons. Husserl writes: "it is the case also of experience that it is never perceived completely, that it cannot be adequately seized upon in its full unity"

[22] "Alles, was wir von der Dinggegebenheit ausgeführt haben, verliert hier seinen Sinn[…]." (Hua III/1, 92)

[23] "*Ein Erlebnis schattet sich nicht ab.*" (Hua III/1, 88)

[24] "*Das Erlebnis* stellt sich, sagten wir, nicht dar." (Hua III/1, 92)

(*Ideas I*, 97).[25] Such is the case because no matter how sharply one might draw a distinction between immanent and transcendent perceptions, one cannot help but must retain an element that both types of perception share: both are *temporal* unities. For example, not only the sunset as it appears to me over the hills, in the midst of the light white clouds, but *my seeing of the sunset as well* has its own unique temporality. Such being the case, *experiencing* can only be given to immanent perception as modified, as what continuously recedes into the past, which the now-point of immanent reflection grasps. And even though this type of incompleteness and imperfection is altogether different from the one characteristic of transcendent perception (it is the sunset, and not my seeing of the sunset, that appears perspectivally), nevertheless, one has to admit that neither experience nor experiencing can be given *adequately* and *fully*.

These qualifications suggest that when Husserl characterizes immanent perception as "absolute" (one should not overlook that Husserl himself employs the quotation marks! [Hua III, 81, 163]), the notion of absoluteness must itself be understood in a relative sense, i.e. as relative to the type of givenness characteristic of transcendent perception.[26] It is this lack of adequacy and fullness, which we encounter at the heart of immanent perception, that provides us with a glimpse of the horizonal nature of experiencing. My *seeing* of the sunset can only retain its unity of sense in each and every now-point by means of its reference to a system of retentions and protentions which must play an irreducible role in each and every act of experiencing. Without these references, no act of consciousness could be an act at all, since it would have to be reduced to an infinite set of "experiencings." Even worse, since these infinitely minute "*Erlebnisse*" would have no unity of their own, they themselves would not be even possible. Thus experiencing is possible only as an already unified process which in its own turn presupposes a horizon of retentions and protentions, due to which each now-point of experiencing can be what it is: *experiencing. Let us call this reference of each now-point of experiencing to its temporal horizon the second sense of the horizon of self-consciousness.*

However, in the context of distinguishing immanent and transcendent perceptions, Husserl does not engage the problem of the unity of experiencing. In *Ideas I*, the question of temporal horizons emerges in a context which, besides securing the unity of particular *Erlebnisse*, also aims to disclose the unity of the stream of consciousness itself. These indeed are not unrelated issues, since the unity of a particular lived-experience is determined not only from "within," but also from "without." That is, just as each object, so each experience also has an inner- and an outer-horizon. Such is the case because the unity of experiencing is secured not only

[25] "Auch ein Erlebnis ist nicht, und niemals, vollständig wahrgenommen, in seiner vollen Einheit ist es adäquat nicht fassbar." (Hua III/1, 93)

[26] This type of "relativization" of the absolute consciousness of *Ideas I* finds its further confirmation in §81, where Husserl remarks that the transcendental "Absolute" (Husserl here again employs this term in quotation marks) that the reductions bring forth is not in fact the last and truthful absolute time-consciousness (Hua III/1, 182).

through the play of retentions and protentions, but also through the immanent object's position within the whole stream of pure consciousness.

Husserl briefly touches upon the latter theme in §57 of *Ideas I*, where he argues that the pure ego, much like Kant's "I think," is a necessary correlate of the stream of lived-experience.[27] The stream, as the phenomenological residuum, is not given as a plurality of different streams, but as one unified field of consciousness. When the question concerning the ground of this unity arises, Husserl finds himself compelled to claim that the stream has an identical pure I as its correlate, whose constant co-presence stamps each lived-experience with the imprint of *mineness*.[28] This belonging of each experience to one and the same stream is not utterly empty of sense; in fact, it faintly hints that the sense of a particular experience is inseparable from the co-presence of other experiences, *a co-presence which functions as a horizon*.[29]

Here we are on the brink of recognizing the unique type of horizon-consciousness distinctive of immanent perception. One can allude to its inimitable character by saying that while transcendent perception schematizes the horizons primarily in terms of space, immanent perception schematizes them exclusively in terms of time. Admittedly, this Kantian language is somewhat misleading, since it seems to suggest that the temporal horizons should be understood as the expression Husserl uses to characterize succession. Yet mere succession is not a figure of a horizon: it does not reveal how our consciousness of the past and the future co-determines our consciousness of the present. We thus need to take a closer look at the temporal horizons of immanent perception.

Each and every *Erlebnis* has a *necessary* horizon of the before, the after, and of the now. As Husserl emphasizes, none of these horizons can be merely empty forms without content. Each experience has a horizon of the before, because "*no experience can cease without there being consciousness of the ceasing and of the having*

[27] Consider also Husserl's remark in Hua XXXIII: "…[w]as dauert, hat in jeder Phase der Dauer einen neuen Gehalt, das Ich aber hat in der Zeit gar keinen Gehalt, nichts Verschiedenes und nichts Gleiches." (Hua XXXIII, 280)

[28] The domain of *mineness*, however, is established as "transcendence within immanence," which needs to be kept in mind when confronted with such an interpretation of the pure ego in *Ideas I* as we find in Shigeru Taguchi's *Das Problem des 'Ur-Ich' bei Edmund Husserl*: "Die neuartige Gegebenheit des reinen Ich muß nun näher charakterisiert werden. Jene 'Unscheinbarkeit' des reinen Ich weist zunächst darauf hin, daß mein Ich für mich selbst als den phänomenologisch Betrachtenden nichts anderes ist als meine 'Nähe' selbst, die ich nicht 'aus der Distanz' beobachten kann." (Taguchi 2006, 68–69) What remains problematic in such an interpretation is the manner in which the "Nähe" is to be understood in relation to time-consciousness. According to Taguchi, "Die zeitliche 'Unerscheinbarkeit' des reinen Ich [resultiert] aus seiner radikalen 'Nähe'" (71). Yet does the recognition of the stream-like character of consciousness not demand one to interpret the "timeless" ego as a functional notion, rather than a genuinely phenomenological finding? Such a functional position is defended by Dieter Lohmar, in the so-far unpublished "The History of the Ego: The 'Arch-Ego' in Husserl's Late Manuscripts on Time and the *Crisis*." I will provide a much more detailed analysis of these themes in Chap. 9.

[29] It is interesting to note that the notion Husserl employs here is not that of *Horizont*, but rather that of *Hintergrund*. "In jedem aktuellen cogito lebt es [das reine Ich] sich in besonderem Sinn aus, aber auch alle Hintergrundserlebnisse gehören zu ihm und es zu ihnen[…]." (Hua III/1, 123)

ceased, and that is a newly filed out Now" (*Ideas I*, 196).[30] Similarly, each experience has a horizon of the after, because each experience is experienced as ceasing and thus as passing into a new now. In the words of William James, "the present image shoots its perspective far before it, irradiating in advance the regions in which lie the thoughts as yet unborn" (James 1950, 256). Finally, each experience has a horizon of the now because each experience does not only fall under the rubric of succession, but also under that of simultaneity, and only along with other lived-experiences which share the originary form of the now does it constitute what Husserl calls the *Originaritätshorizont des reinen Ich.*

This implicit reference that each lived-experience draws to the ones that precede and follow it does not merely play an irreducible role in securing the identity of a particular *Erlebnis*. It is of no less importance that this constant play of references has its limits: *the outermost horizon of these implicated allusions is nothing other than the stream of consciousness itself. Let us call it the third sense of the horizon of self-consciousness*. It is the outermost horizon of sense, because what lies beyond it—the experiences given within other streams of consciousness—do not perform a constitutive role in obtaining the sense of a particular experience, which unfolds within my field of consciousness. Thus Husserl speaks of the stream of consciousness as the *temporal field of the pure I* (*Zeitfeld des reinen Ich*) and, what one must not overlook, of the "stream of consciousness summoning itself in its continuity of content" ("inhaltliche Kontinuität fordernder Erlebnisstrom" [*Ideas I*, 196, Hua III/1, 165]). The unity of the stream cannot be empty or formal since it is engendered through the implicit references, which secure the sense and identity of each particular *Erlebnis*. Thus the horizon of a particular lived-experience is something altogether different from the horizon of the temporal field of the pure I; and yet, both are continuously engendered through the same system of references which accompany each act of experiencing.

With the emergence of the horizon of the stream of lived-experiences, we face a new dimension of horizon-consciousness: although the stream itself never appears as actual, nevertheless, it is continuously given to consciousness. It never appears as actual, because it can never manifest itself as an *Erlebnis* among other *Erlebnisse*, which means that *it is given only horizonally*. Each and every *Erlebnis* is a *dependent* moment of the *Erlebnisstrom*; and since the latter is not a moment of itself, it cannot appear as its own moment, i.e., *it can only appear as a horizon*. And yet, despite its *actual* emptiness, it has an irreducible content of sense, which in its infinite extension is given, and arguably, given indubitably.[31]

[30] "*Kein Erlebnis kann aufhören ohne Bewußtsein des Aufhörens und Aufgehörthabens*, und das ist ein neues ausgefülltes Jetzt." (Hua III/1, 184)

[31] Its indubitability is of a different sort than the indubitability that pertains to the givenness of what is immanently perceived. The latter is indubitable, because it does not manifest itself through "*Erscheinungsweisen*," but is rather given as "it itself" in a simple grasp. The former admits of the possibility of error, since my grasp of the experience that lies in the past or the future can be mistaken. And yet, the recognition of this error is itself possible only on the basis of the pregivenness of the stream of lived-experiences. Far from denying the unity of *Erlebnisse*, the possibility of error pertaining to an actual experience only confirms it.

This is an important realization, for there is a common tendency to associate the horizon with the background/foreground schema and to assume that this schema is exhaustive of the horizon-problematic. As we now see, the stream of consciousness constitutes a notable exception. And as we will see in Chap. 10, it is not the only exception: there is a sense in which the world-horizon is also given in the absence of any intuitive core.

5.7 The Limits of Husserl's Early Analysis of the Horizon

As we have seen, *Ideas I* analyzes the horizon as an essentially two-sided phenomenon, which embraces not only the objects of the acts of consciousness (i.e., appearances), but also the acts themselves (i.e., experiencing of these appearances). Yet in *Ideas I* this two-sidedness, besides being exposed, is simultaneously restricted from revealing its far-reaching significance. And thus, for the reasons I will turn to in this section, Husserl's early analysis of the horizon remains significantly curtailed and deficient.

The central reason why Husserl's early analysis of the horizon is incomplete consists in the fact that this analysis misses the phenomenon of the world. That such is the case, we can already glimpse in the surprising lack of parallelism between the horizons of the acts of consciousness and the horizons of the objects that these acts intend. At the end of the last section, my analysis of the horizonality of experiencing culminated in the figure of the stream of consciousness which revealed itself as horizonal, even though it lacked any kind of intuitive core. Yet on the side of appearances, a parallel horizon-figure was not drawn. Why? Arguably, because its establishment would require that Husserl's analysis, besides thematizing the objects of the conscious acts, would also address the world-horizon. Yet as we have seen in Chap. 4, *Ideas I* does not problematize the horizon at this level. Due to this lack of parallelism, the horizon can be characterized as a theme whose phenomenality in *Ideas I* is not only uncovered but also suppressed, revealed, yet not effectively resolved.

Yet as we have also seen in Chap. 4, the resources for establishing the parallelism in question are already present in *Ideas I*. After all, the first figure of the horizon that we find in this text is precisely that of the world-horizon. We thus need to ask what proscribes phenomenology from a more probing analysis of the horizons.

Here we are led to a paradoxical conclusion: *what establishes the conceptual space within which the horizon reveals itself in its phenomenality is exactly what suppresses its full manifestation.* I am referring to the *reduction* as thematized in *Ideas I*. In this work, much like in static phenomenology more generally, the phenomenological analysis starts with objects, be they real or ideal, and leads their meaning back to their establishment in the acts of consciousness. Such an analysis does indeed clarify the sense of objectivity by unearthing those structures in virtue of which objects get to be constituted in consciousness. *Yet the world is not an object.* So how could it have a prominent place within such a phenomenological framework? The horizons that get to be uncovered are object-horizons as well as horizons of consciousness, i.e., horizons of those structures of consciousness in virtue of which objects can become themes of consciousness.

There is, however, yet another reason why Husserl's early analysis of the horizon remains inconclusive. In *Ideas I*, even Husserl's analysis of the horizons of consciousness has its own limits. So as to see how unprecedented, yet simultaneously restrictive, Husserl's discovery of the horizon is, let us return to Husserl's claim that intentionality is phenomenology's main theme—a position Husserl argues for in §84. On its initially superficial level, intentionality stands for the correlative bond between perceiving and the perceived, judging and the judged, valuing and the valued, etc. Yet to this Husserl adds that the explicit cogito is only "a particular modality of that something universal which we call intentionality" (*Ideas I*, 201).[32] The *potential* experiences and the *potential* appearances are just as strongly tied with an intentional bond; they also are inseparable correlates. Moreover, these background-intentions and horizonal appearances can manifest themselves on a number of different levels, depending on their proximity to or distance from the actual cogito and actual appearance. The task of clarifying intentionality—"phenomenology's main theme!"—thereby overlaps with the clarification of the horizonality of experience.

Such is the message of §84, yet its actual sense seems to be suppressed as soon as it is announced. For what the recognition of intentionality as phenomenology's main theme leads to is not the uncovering of the different dimensions of sense which are irreducible despite their inability to appear on the level of the explicit cogito. Rather, in *Ideas I* intentionality is questioned not in terms of its "*stirrings*" (*Gefallensregungen*, *Urteilsregungen*, and *Wunschregungen* which Husserl mentions in §84), but rather in terms of its immobile elements, such as *hyle*, *noesis*, and *noema*, which remain what they are whether or not we thematize the actual or the potential cogito. To borrow a metaphor from William James, just as a snowflake caught in a warm hand is no longer a crystal but a drop, so the horizon reduced to its essential elements is no longer what it once was before. The depth of intentionality is thereby covered up as soon as it is discovered. To continue with Jamesian metaphors, one could say that *Ideas I* talks of intentionality like one who should say "a river consists of nothing but pailsful, spoonsful, quartpotsful, barrelsful, and other moulded forms of water. Even were the pails and the pots all actually standing in the stream, still between them the free water would continue to flow" (James 1950, 255). It is just this free water that Husserl discovered, yet resolutely overlooked, in *Ideas I*. For indeed, if the metaphor of the stream is to have any sense in characterizing the field of *Erlebnisse*, then the attempt to reveal intentionality in terms of its static elements must have clearly circumscribed limits which one sooner or later needs to transgress so as to unravel what lies hidden in intentional consciousness. What makes the stream of consciousness into a stream remained unanswered in *Ideas I*, even though the necessary elements for addressing this issue were already disclosed in this work.

[32] "[…] vielmehr galt uns dieses Cogitative als eine besondere Modalität des Allgemeinen, das wir Intentionalität nennen." (Hua III/1, 189)

Thus the analysis of the horizon in *Ideas I* can be likened to the painter's first sketch on the canvas. It is as if the contours of the horizon are drawn with a shaking hand, still uncertain of what is to come out of the drawing. Or better, it is the free movement of the hand that is missing and hence the figure of the horizon is drawn as if composed of infinitesimally small straight lines thereby concealing the phenomenon's fluid appearance. By the time the notions of noema and noesis occupy the center of phenomenology's attention—and these sections are undoubtedly the most important ones in *Ideas I*—we no longer find horizonal investigations.[33] We are now in the position to see why such is the case. Characterizing intentionality as phenomenology's main theme and horizonality as its fundamental feature (§84), phenomenology finds itself at a crossroad. On the one hand, it feels compelled to transcendentally ground the possibility of object-consciousness (and thus it investigates the object as such), reveal its universal features, in short, generate a theory of the noema. This is the path *Ideas I* chooses as soon as it enacts the "transcendental turn" in §86. But on the other hand, if it chooses this path, it overlooks precisely what makes it possible—the phenomenality of the horizon itself, whose analysis requires one to trace the main lines of the subjective context which is hidden from the objective view, unearth the vestiges of sense which stamp each and every object as part of the already human environment.[34] Being limited from the start to the analysis of the structures which pertain to the elements of lived-experience, no matter whether the latter are actual or potential, the analysis undertaken in *Ideas I* no longer finds the motivating force to continue with the horizonal investigations. As we will see, Husserl's further elaboration of horizon-consciousness signifies the revival of what was already alluded to in §84, viz., the revival of the task to uncover the intentional framework of consciousness at the diverse levels of the potential cogito—the levels that remain concealed from the analysis directed at the structures of the acts of consciousness. Husserl's further reaching analysis will take place within the context he qualifies as that of the "hidden subjectivity."

To recapitulate: why is the horizon not revealed in its full phenomenality in the text, which marks the discovery of the horizon in philosophy? First, because *Ideas I* misses the phenomenon of the world; and secondly, because it also neglect the horizon of subjectivity. And why is the world-horizon and the horizon of subjectivity not disclosed in the confines of static phenomenology? For all intents and purposes, because Husserl's early way to the reduction proscribes their further-reaching disclosure.

---

[33] It is telling that the few times the notion of the horizon emerges in these parts of the text, it is used in the most imprecise sense, i.e., merely as it functions in everyday speech, as for instance *Problemhorizont* (§146). Such casual employment of this notion shows particularly clearly (even clearer than it would had the notion of the horizon not been employed at all) that the genuine sense of the horizon-consciousness is no longer significant for the central goals of *Ideas I*. When the notion emerges in its phenomenological sense (e.g., Hua III/1, 327), it does not enrich what we already know from its earlier treatment.

[34] See Kuhn (1940, 121).

The subsequent analysis thereby obtains a clear determination. First, it will be essential to show the difference between the static and the genetic notions of the horizon (Chap. 6). Secondly, it will be also crucial to address the problematic of the reduction so as to obtain the means necessary to inquire into more hidden aspects of intentionality (Chap. 7). Afterwards, we will be in the position to turn to the analysis of the horizon of subjectivity (Chap. 8). Finally, after bringing such a genetic notion of the horizon into a dialogue with hermeneutical accounts of the horizonality of experience (Chap. 9), we will bring our analysis to completion by turning to the problematic that falls under the heading of the world-horizon (Chaps. 10, 11 and 12).

Having reached the end of the first part of this work, let us ask: *what is consciousness conscious of when it confronts an appearing objectivity*? And correlatively, *what is it that appears in the appearance of the given objectivity*? At the outset one might have been tempted to respond: but of course, what appears is the very object that consciousness confronts, and it is this very object that is the terminus of the conscious intention. Now we see why such a response is insufficient. The phenomenological reflection undertaken in the first part of this work shows that no "being in itself" or "subject in itself" can be presupposed in accounting for horizon-consciousness, for it is within these horizonal structures that the very sense of being and of subjectivity originates. Both the appearance and the conscious intention have manifold levels of sense without which no appearance and no intention are possible. The object appears in terms of its inner- and outer-horizons; the intention is actual only in the context of an almost infinite set of tacit intentions. The object is what it is only because it appears in the framework of non-actual appearances; the intention is what it is only because it emerges within the framework of other past, future, and present intentions. We also see that the uncovering of the tacit dimensions of intendings is by far not unrelated to the exposure of the non-actual dimensions of appearances. Rather, the two are correlated with each other within an intentional framework. We thereby see the unavoidable dependence of actuality upon potentiality. Yet as we acknowledge the correlative nature of horizonal investigations, we simultaneously need to concede a certain lack of parallelism between the analysis undertaken on the level of experience and of experiencing. Overcoming this shortcoming will reveal what *Ideas I* already hint at, viz., that far from restricting the intending of an object and the appearing of its appearance to the object's manifestation, the horizon-intentionality co-intends the pregivenness of the world. However, before we turn to the world-horizon, we first need to see what it means to qualify the horizon as a genetic theme and a genetic phenomenon.

References

Brenner, Andreas. 2006. Non-intentionality and the lived-body. In *Analecta Husserliana*, vol. LXXXIX, ed. Anna-Teresa Tymieniecka. Dordrecht: Springer.

Cairns, Dorion. 1976. *Conversations with Husserl and Fink*. The Hague: Martinus Nijhoff.

Claesges, Ulrich. 1964. *Edmund Husserl's Theorie der Raumkonstitution*. The Hague: Martinus Nijhoff.

Husserl, Edmund. 1952. In *Ideen zu einer reinen Phänomenologie und phänomenologischen Philosophie. Zweites Buch: Phänomenologische Untersuchungen zur Konstitution*, ed. Marly Biemel. The Hague: Martinus Nijhoff.

Husserl, Edmund. 1971. In *Ideen zu einer reinen Phänomenologie und phänomenologischen Philosophie. Drittes Buch: Die Phänomenologie und die Fundamente der Wissenschaften*, ed. Marly Biemel. The Hague: Martinus Nijhoff.

Husserl, Edmund. 1976a. In *Ideen zu einer reinen Phänomenologie und phänomenologischen Philosophie. Erstes Buch: Allgemeine Einführung in die reine Phänomenologie*, ed. Karl Schuhmann. The Hague: Martinus Nijhoff.

Husserl, Edmund. 1976b. In *Die Krisis der europäischen Wissenschaften und die transzendentale Phänomenologie. Eine Einleitung in die phänomenologische Philosophie*, ed. Walter Biemel. The Hague: Martinus Nijhoff.

Husserl, Edmund. 1983. *Ideas pertaining to a pure phenomenology and to a phenomenological philosophy. First book: General introduction to a pure phenomenology.* Trans. F. Kersten. The Hague: Martinus Nijhoff.

Husserl, Edmund. 1999. In *Erfahrung und Urteil. Untersuchungen zur Genealogie der Logik*, ed. Ludwig Landgrebe. Hamburg: Felix Meiner Verlag.

Husserl, Edmund. 2001. In *Die 'Bernauer Manuskripte' über das Zeitbewußtsein (1917–1918)*, ed. Rudolf Bernet and Dieter Lohmar. Dordrecht: Kluwer Academic Publishers.

James, William. 1950. *Principles of psychology*, vol. I. New York: Dover Publications.

Kuhn, Helmut. 1940. The phenomenological concept of the 'horizon'. In *Philosophical essays in memory of Edmund Husserl*, ed. Marvin Farber. Cambridge, MA: Harvard University Press.

Landgrebe, Ludwig. 1973. The phenomenological concept of experience. *Philosophy and Phenomenological Research* 34(1): 1–13.

Pietersma, Henry. 1973. Intuition and horizon in the philosophy of Husserl. *Philosophy and Phenomenological Research* XXXIV: 95–101.

Taguchi, Shigeru. 2006. *Das Problem des 'Ur-Ich' bei Edmund Husserl. Die Frage nach der selbstverständlichen Nähe des Selbst.* Dordrecht: Springer.

Part II
The Horizons of Transcendental Subjectivity

Chapter 6
The Static and Genetic Determinations of the Horizon

6.1 Static and Genetic Phenomenology: A Preliminary Account

The central task of this chapter is that of showing that as a genetic notion, the horizon is a horizon of subjectivity. So as to establish this thesis, I will begin by drawing a methodological distinction between static and genetic phenomenology. On this preliminary basis, I will turn to Husserl's general qualification of the horizon as a horizon of consciousness. I will subject this notion to a static and genetic analysis, thereby exemplifying the difference between the methods in question. Such a procedure will generate the insight that the horizon is a distinctly genetic phenomenon. So as to obtain a better sense of its phenomenality, I will turn to the different notions of the ego in Husserl's phenomenology. We will thereby come to see that as a genetic phenomenon, the horizon is a horizon of subjectivity.

Although I have early on alluded to the horizon as a genetic phenomenon, I have so far left the meaning of this characterization largely undetermined. The next three chapters aim to provide this characterization with a more precise determination. However, even before one inquires into static and genetic determinations of the horizon, at least a preliminary account of the difference between static and genetic phenomenological methods is called for. Let me emphasize from the start that my subsequent analysis in this section is meant to be only *preliminary*. It is my contention that our understanding of the difference between static and genetic phenomenology can be more successfully established not by tackling the methodological issues in a straightforward way, but rather by turning to specific phenomena and, on the basis of their description, exemplifying the difference between the methods in question. In the subsequent sections, the phenomenon of the horizon will serve as an example.

Although one can find genetic elements in Husserl's analysis already in the *Logical Investigations*, according to Husserl himself, his explicitly genetic analyses reach back to around 1917. It was only around 1921 that Husserl himself was led to

S. Geniusas, *The Origins of the Horizon in Husserl's Phenomenology*, Contributions to Phenomenology 67, DOI 10.1007/978-94-007-4644-2_6,

formulate an account of the difference between the static and genetic methods. It is interesting to note that even though Husserl qualifies his pre-1917 phenomenology as static, the very notion, "static phenomenology," does not make its appearance until the early 1920s, when it was placed in opposition to genetic phenomenology. Husserl's own clarification of static and genetic methods notwithstanding, the difference between these phenomenological methods is notoriously vague and the plurality of interpretations that address them did not generate any kind of consensus. In this section, I will primarily build my provisional account of the difference between the methods in question on the basis of Husserl's short text that he wrote in 1921, significantly titled "On Static and Genetic Phenomenological Methods."[1]

The very notions, "stasis" and "genesis," invite one to claim that while static phenomenology is primarily concerned with stable structures of experience, genetic phenomenology raises the question of their origination. Although such a characterization is not incorrect, it nonetheless remains too general, and for this reason, it can all-too-easily lead to misunderstandings. Consider in this regard Mary Jeanne Larrabee's "Husserl's Static and Genetic Phenomenology." As Larrabee puts it,

> a simple description of static analysis is that it analyzes a constituted object without dealing with the flowing temporal process in which the object is constituted; it considers the object's constitution as though it were halted at its termination. For this reason, static analysis is commonly characterized as lacking all reference to time. (Larrabee 1976, 163–164)

And yet, while Husserl himself declares that his phenomenological method has been primarily static until about 1917, it nonetheless remains the case that the hundreds of pages of Husserl's analysis of time-constitution, compiled in *Husserliana X*, span the years from 1893 until 1917. This elementary fact necessitates one to admit that the problematic of time does not lie outside the scope of static analysis. In its own turn, this admission calls for a more refined distinction between static and genetic phenomenological methods.

The distinction in question is not thematic but rather methodological. By this I mean that one cannot draw a distinction between static and genetic phenomenology on the basis of certain themes being allegedly addressed in one and being absent from the other. Rather, any theme whatsoever—and the problematic of time does not constitute an exception—can be addressed both from a static and from a genetic perspective.[2]

Given the misunderstanding mentioned above, it is interesting to observe that Husserl, in the text from 1921 to which I have referred above, begins his analysis with the claim that the phenomenological understanding of the laws of genesis

[1] See Hua XI, pp. 337–345.

[2] However, even though the distinction between static and genetic phenomenology is methodological, it nonetheless remains the case that some themes are better treated in a static, while others in a genetic perspective. The problematic of the horizon is a good case in point: one could qualify it (as I have done above) as a genetic phenomenon, but not because there are no static analyses of the horizon. Rather, such a qualification is meant to suggest that a static analysis of the horizon is significantly limited and that a genetic treatment of the horizon is capable of overcoming these limits.

can be understood at least in two different ways. On the one hand, by the laws of genesis, one can understand those laws that rule over the *emergence of "events" in the stream of lived-experiences*, i.e. laws that rule over the *connections of retentions, impressions, and protentions*. This means that Husserl begins his own account of the distinction between static and genetic methods precisely with an admission that the question of genesis itself does not lie outside the scope of static phenomenology, and thus, it cannot be said in an unqualified way that static phenomenology lacks any kind of reference to time. Of course, this should not be taken to mean that the problematic of time-constitution belongs exclusively to static phenomenology. Rather, the first notion of the laws of genesis suggests that the question of genesis can be addressed both statically and genetically. On the other hand, by the laws of genesis one can also understand those laws that rule over the *formation of apperceptions*. As we will soon see, just as time-constitution, so the problematic of apperceptions also, can be addressed both statically and genetically. Husserl's analysis in the text from 1921 thereby confirms the claim that the distinction between static and genetic phenomenology is not thematic, but only methodological.

Yet even though the distinction between static and genetic phenomenology is methodological, it can be best grasped not on a plainly methodological level, but rather within a thematic context that would exemplify this distinction. In the text under consideration—one of the very few in which Husserl thematizes the distinction between static and genetic methods—Husserl exemplifies the difference between the methods in question by turning to one particular theme, viz., that of *apperceptions*.

By apperception, Husserl understands such a consciousness, which is not merely conscious of something, but also intends this something as a *motivation* for being conscious of something else.[3] The problematic of apperceptions thereby proves to be inseparable from that of the horizon. Bringing the two notions into proximity with each other, one could say that every kind of consciousness carries with it its own *apperceptive horizons of sense*, and thus, the problematic of apperceptions characterizes not just one particular type of consciousness among others, but rather *consciousness in its entirety*. For instance, as my analysis in Chap. 5 has shown in detail, in the case of perception, the intention directed to a particular profile of an object is always co-conscious of the other ways in which this very object can be given. We encounter this very same structure in all other types of consciousness. With this in mind, Husserl asserts:

> apperception is a concept that encompasses every self-giving, thus every intuitive consciousness .... In fact, we cannot even conceive of a consciousness that would not go beyond the strict present in its essential flux from presence to new presences; consciousness is

---

[3] "Könnten wir nicht auch Apperzeption so definieren: ein Bewußtsein, das nicht nur überhaupt etwas in sich bewußt hat, sondern es zugleich als Motivanten für ein anderes bewußt hat, das also nicht bloß etwas bewußt hat und zudem noch ein anderes darin nicht Beschlossenes, sondern das auf dieses andere hinweist als ein zu ihm Gehöriges, durch es Motiviertes." (Hua XI, 338)

> inconceivable without retentional and protentional horizons, without a co-consciousness (although necessarily a non-intuitive one) of the past of consciousness and an anticipation of an approaching consciousness (no matter how indeterminate it may be) (APS, 625–626).[4]

Since consciousness without apperceptive horizons is inconceivable, one can assert, along with Husserl, that *the universal doctrine of consciousness is the universal doctrine of apperceptions* (Hua XI, 340). With this in mind, let us take a look at how the problematic of apperceptions arises in the context of static and genetic phenomenologies.

How exactly does the so-called static phenomenology thematize apperceptions? One could answer this question by saying that static phenomenology addresses apperceptions *structurally* and *teleologically*. Here the term "structural" refers to phenomenology's aim to give an account of the essential structures of consciousness, while the term "teleological" designates different modes of fulfillment, to which intentions are subject. Static phenomenology aims to show how each apperception exhibits a noetic-noematic structure. Furthermore, it also gives an account of the different ways in which apperceptions carry out sense-giving as well as how they posit objects in one doxic modality or another. Finally, one does not overstep the boundaries of static phenomenology when one inquires into the different types of fulfillment to which apperceptions can lead, or when one describes these modes of fulfillment noetically and noematically.

Such a structural and teleological account of apperceptions, be it provided in the context of intentions and their fulfillment, or in the context of analyzing the eidetic structures of consciousness, does not give rise to an understanding of the *genesis* of apperceptions. For this reason Husserl argues that even though static phenomenology addresses apperceptions in the above-mentioned way, it nonetheless fails to inquire into their genesis. In Husserl's words, static phenomenology deals with "finished" apperceptions. Such a characterization intimates that there must be a certain "history" that underlies the "finished" manifestation of apperceptions. The task of genetic phenomenology is precisely that of uncovering this "history."

The "history" in question is unusual (hence the quotation marks) in that it is not meant to be factual, but rather *eidetic*. In Husserl's words, every shape of apperception is an *essential* shape, and thus, one can study its genesis according to *essential necessity*. The question regarding the genesis of apperceptions opens the gate to genetic phenomenology.

How are we to understand such an eidetic "history?" First of all, one must note that its possibility relies upon the hypothesis that that the genesis of particular apperceptive horizons within the stream of lived-experiences is ruled by universal

---

[4] "Apperzeption […] ist ein Begriff, der jedes selbstgebende, jedes anschauliche Bewußtsein also umspannt[…]. In der Tat ist keines denkbar, das in seinem wesensmäßigen Fluß von Präsenz zu neuen Präsenzen nicht über das eigentlich Präsente hinausgriffe, keines ist denkbar ohne retentionale und protentionale Horizonte, ohne ein Mitbewußtsein (obschon ein notwendig unanschauliches) der Bewußtseinsvergangenheit und eine Vorerwartung von kommendem Bewußtsein (wie unbestimmt es auch sein mag)." (Hua XI, 337–338)

necessity (Hua XI, 338). Such a hypothesis is not unwarranted insofar as it finds experiential confirmation, i.e., insofar as we can locate in experience necessary laws that necessitate the emergence of particular apperceptions, and Husserl is confident that genetic phenomenology can justify such a hypothesis. One of the central tasks of genetic phenomenology is precisely that of disclosing such a universal necessity. As Husserl puts it, "it is a necessary task to establish the universal and primitive laws under which stands the formation of an apperception arising from a primordial apperception, and to derive systematically the possible formations, that is, to clarify every given structure according to its origins" (APS, 627).[5]

How does genetic phenomenology address the genesis of apperceptions? It addresses this issue on three different levels of experience, which in a regressive way, one can describe as follows. First, genetic phenomenology inquires into the formation of apperceptions on the level of *pure activity*, i.e., the level at which consciousness faces ideal objects and accomplishes their real generation. In this context, the problematic of habituality plays a crucial role. Secondly, genetic phenomenology also thematizes the formation of apperception at the deeper level of the *intersection between activity and passivity*. Here the inquiry into the preconditions of the formation of ideal objectivities proves crucial. Thirdly, genetic phenomenology also directs its attention at the genesis of *passivity* and the general lawful regularity of *purely passive becoming*. At this level, the genetic analysis of time-constitution obtains a central role.

On the basis of the foregoing analysis, one can formulate the difference between static and genetic phenomenology in the following way. When phenomenology employs the static method, it begins with stable intentional objects, be they real (natural things) or ideal (logical or mathematical propositions). Approaching these objects as functions of sense-giving, static phenomenology aims to accomplish a twofold task: show *how* objects are constituted, as well as account for the essential structures that enable their constitution. This emphasis on the modes of givenness of objectivities and on the essential structures of consciousness allows one to qualify static phenomenology as constitutive. By contrast, when phenomenology employs the genetic method, it raises the question of the *origins of such constitutive frameworks*, which bind experience to objectivity. That is, *the genesis that genetic phenomenology aims to uncover is precisely the genesis of constitution*. For this reason, in *Formal and Transcendental Logic*, Husserl asserts that

> (inquiry into) the *"static" constitution* of objects, which related to an already "developed" subjectivity, has its counterpart in (an inquiry into) *a priori genetic constitution*, (a subsequent inquiry), based on (the results of) the former, which necessarily precedes it.[6] (FTL, 250)

[5] "Es ist also eine notwendige Aufgabe, die allgemeinen und primitiven Gesetze, unter denen die Bildung von Apperzeption aus Urapperzeption steht, aufzustellen und die möglichen Bildungen systematisch abzuleiten, also jedes gegebene Gebilde seinem Ursprung nach aufzuklären." (Hua XI, 339)

[6] "Dann entspricht der '*statischen*', auf eine schon 'entwickelte' Subjektivität bezogenen Konstitution von Gegenständen die apriorische genetische Konstitution, aufgestuft auf jener notwendig vorangehenden." (Hua XVII, 257)

Thus while static phenomenology aims to account for the already developed and, as Husserl sometimes says, "finished" constitutive systems by inquiring into intentional experiences as they fall under eidetic laws, the task of genetic phenomenology is that of uncovering the origins of these very systems. The "history" that genetic phenomenology uncovers thereby proves to be both the "history" of subjectivity and of the object. It is the "history" which spans the essential interweaving of different shapes of consciousness, a "history" that shows how a particular shape of consciousness must arise from other shapes of consciousness, thereby leading to the "finished" constitutive systems, which in their own turn constitute the starting points of static phenomenology.

As mentioned above, this account of the difference between static and genetic methods is meant to be only preliminary. With it in mind, let us now turn directly to the horizon-problematic and see how it calls for a genetic treatment. For if we ask, what exactly motivates Husserl to enlarge his so-called static analyses with genetic investigations, the answer that would point in the direction of Husserl's indebtedness to Neo-Kantians[7] or Dilthey[8] would remain severely constrained. First and foremost, it is phenomena themselves, as they emerge in the context of static phenomenology, that motivate Husserl to pursue genetic investigations.[9] The horizon is precisely such a phenomenon; or as I have argued in the first part of this investigation, the horizon is a genetic phenomenon, which in its first appearance is still dressed in static garb.

Yet how are we to reconcile the claim that the horizon is a genetic phenomenon with my foregoing suggestion that the distinction between static and genetic phenomenology is not thematic, but rather methodological? On the one hand, the first of these claims invites one to admit that in fact, the distinction between static and genetic phenomenology is not just methodological, but also thematic. On the other hand, the second of these claims precisely invites one to give up the prospect of drawing a distinction between static and genetic phenomenology on a thematic basis.

Yet these two claims only appear to be irreconcilable. On the one hand, the distinction between static and genetic phenomenology is methodological, and *only* methodological. On the other hand, this does not mean that both methods are of equal significance for the analysis of any phenomenon. Different phenomena

[7] In a letter to Paul Natorp, written in June, 1918, Husserl observes that "already for more than a decade, I have overcome the level of static Platonism and have situated the idea of transcendental genesis in phenomenology as its main theme."

[8] In the manuscript under consideration, Husserl refers to the distinction between static and genetic phenomenology in the same terms that Dilthey used for psychology, viz. "descriptive" and "explanatory." However, these notions mean something different for Husserl and for Dilthey. For Dilthey, description is closely tied to interpretation, whereas explanation is closely tied to those procedures, which we find in the natural sciences. The foregoing analysis must have made clear that Husserl understands this distinction quite differently.

[9] Or as A. Steinbock puts it, "the phenomenal field can on its own part overstep the bounds of a pronounced or presupposed methodological undertaking and demand the formulation of a new methodology." (Steinbock 1998, 127)

can be treated better or worse by each of these methods. Thus my claim that the horizon is a genetic phenomenon need not be understood as a suggestion that a static account misses the phenomenon of the horizon. To qualify the horizon as a genetic phenomenon is to suggest that the genetic method has the means of generating a more embracive account of the horizon. So as to see that such indeed is the case, let us turn to Husserl's notion of horizon-consciousness. Section 6.2 will show what is distinctive about Husserl's notion of horizon-consciousness, while Sects. 6.3 and 6.4 will further show how this notion is treated in static and genetic phenomenology.

6.2 A General Determination of the Horizon as Horizon-Consciousness

Arguably, the most distinctive trait of Husserl's analysis of the horizon consists in the fact that Husserl thematizes horizon-intentionality as the *horizon of consciousness*. Such a manner of inquiring into the horizon appears questionable as soon as one takes into account how the horizon-problematic has been developed in post-Husserlian thought. On the one hand, the claim that the horizon is a horizon of consciousness seems to be nothing more than an indication of the most *general* trait that qualifies the manner in which the horizon continues to be analyzed in phenomenology and hermeneutics. This notion bespeaks the *subjective reference* that each and every horizon entails, and the question how this reference is to be understood continues to dominate the philosophical analyses of the horizon. Yet on the other hand, the identification of the horizon as horizon-consciousness also suggests something *distinctive* about Husserl's manner of investigating the horizon. For Husserl, to qualify the horizon as the horizon of consciousness is to suggest that *the horizon is the horizon of the ego*. Husserl further interprets this egological dimension as the horizon of *transcendental subjectivity*, through whose *constitutive* accomplishments the horizons come to be what they are. Thus for Husserl, *the horizon's subjective reference implies the horizon's constitutive and transcendental dimensions*—dimensions which on a variety of fronts are challenged not only by Husserl's adversaries, but also by his followers. Horizon-consciousness thereby entails not only what is appropriated, but also what is rejected in post-Husserlian philosophy.

A brief look at Hans-Georg Gadamer's *Truth and Method* will enable us to corroborate both claims. In his magnum opus, Gadamer does not hide that he appropriates the notion of the horizon from Husserl and Nietzsche[10] and introduces it as a central characteristic of what he calls "consciousness of being affected by history"

[10] "Insbesondere hat der philosophische Sprachgebrauch seit Nietzsche und Husserl das Wort [Horizont] verwendet, um die Gebundenheit des Denkens an seine endliche Bestimmtheit und das Schrittgesetz der Erweiterung des Gesichtskreises dadurch zu charakterisieren." (Gadamer 1990, 307) See Sect. 1.2 for a brief discussion of the role that the horizon plays in Nietzsche's works.

("wirkungsgeschichtliches *Bewußtsein*").[11] Yet a closer look at the problematic of the horizon reveals that the manner in which the above-mentioned subjective reference is interpreted is precisely where Gadamer and Husserl part ways. Gadamer's critique of the notion of life in Husserl's phenomenology[12] aims to establish an irreconcilable tension between two phenomenological tendencies he finds central in Husserl's thought: the *descriptive* tendency, which depicts the overwhelming presence of the horizons, and the *epistemological* tendency, which manifests itself in the transcendental problematic. In direct contrast to Husserl, Gadamer asserts that to affirm the former means to reject the latter, and therefore the notion of *facticity*, as thematized by Heidegger in *Being and Time*,[13] is central to Gadamer's own hermeneutics. For Gadamer, facticity indicates the rudimentary belonging of the interpreter to the interpreted object, and he insists that it can only be understood in opposition to Husserlian phenomenology. So as to emphasize what he takes to be a central difference between transcendental and hermeneutical phenomenologies, Gadamer writes: "the main point of the hermeneutics of facticity and its contrast with the transcendental constitution research of Husserl's phenomenology was that no freely chosen relation toward one's own being can get behind the facticity of this being" (Gadamer 1989, 264).[14] And thus, in direct contrast to Husserl's inquiry into the horizon as the horizon of transcendental subjectivity, Gadamer proclaims: "the focus of subjectivity is a distorting mirror. The self-reflection of the individual is only a flickering in the closed circuits of historical life" (Gadamer 1989, 276).[15]

In Chap. 9, I will turn to a more detailed inquiry into the hermeneutical critique of phenomenology. At the moment, a brief excursus to the problematic relation between Husserl's and Gadamer's analyses of the horizon is only meant to qualify what the question pertaining to the characterization of the horizon as horizon-consciousness amounts to. This brief excursus shows that to identify the horizon as horizon-consciousness is to address what is *general* to the philosophical analysis of the horizon and simultaneously to question what is *unique* about Husserl's manner of engaging this theme. Thus by addressing this question, I will first of all aim to uncover those dimensions of sense of horizon-intentionality, which call for an inquiry into the horizon as *Horizontbewußtsein*. I hope thereby to show why

[11] See Gadamer (1990), pp. 305–312.

[12] See Gadamer (1990), pp. 246–258.

[13] "Und doch ist die 'Tatsächlichkeit' der Tatsache des eigenen Daseins ontologisch grundverschieden vom tatsächlichen Vorkommen einer Gesteinsart. Die Tatsächlichkeit des Faktums Dasein, als welches jeweilig jedes Dasein ist, nennen wir seine *Faktizität*[…]. Der Begriff der Faktizität beschließt in sich: das In-der-Welt-sein eines "innerweltlichen" Seienden, so zwar, daß sich dieses Seiende verstehen kann als in seinem "Geschick" verhaftet mit dem Sein des Seienden, das ihm innerhalb seiner eigenen Welt begegnet." (Heidegger 2006, 56)

[14] "Daß alles freie Sichverhalten zu seinem Sein hinter die Faktizität dieses Seins nicht zurück kann, darin lag die Pointe der Hermeneutik der Faktizität und ihr Gegensatz zu der transzendentalen Konstitutionsforschung der Husserlschen Phänomenologie." (Gadamer 1990, 268–269)

[15] "Der Fokus der Subjektivität ist ein Zerrspiegel. Die Selbstbesinnung des Individuums ist nur ein Flackern im geschlossenen Stromkreis des geschichtlichen Lebens." (Gadamer 1990, 281)

consciousness is an irreducible dimension of horizon-intentionality; that the decision to divorce the horizon from its bond to consciousness would curtail the horizon from manifesting its distinctive features. Moreover, I will also aim to extract from this general characterization the motives that lead Husserlian phenomenology to thematize the horizon in a *constitutive* framework and to speak of it as the horizon of *transcendental* subjectivity. In contrast to the widespread assumption that this Husserlian motif only suppresses the phenomenality of the horizon, I will argue that the transcendental framework brings to light a crucial and irreducible dimension of the horizon-problematic. Its full significance will become apparent in the final sections of Chap. 7, where I will return to the question of the origins of the horizon in a manner decidedly different from the one that was operative in the first part of this work.

It will thereby become apparent that the exclusion of the horizon's transcendental problematic is nothing less than the exclusion of the possibility to inquiry into the origins of the horizon. Let me emphasize already now that upon this point rests my defense of Husserl's "transcendentalism": it enables one to tie the horizon to the question of origins with an inseparable bond; it even enables one to inquire into the origins of the horizons themselves. As we will see, it enables one to pose questions that in post-Husserlian phenomenology and hermeneutics are already illegitimate—to question those dimensions of horizonality's sense, which within other philosophical frameworks are destined to remain non-phenomenal and non-thematic. Yet before justifying these claims, one first needs address Husserl's notion of *Horizontbewußtsein* in some detail.

6.3 The Static Notion of Horizon-Consciousness

Although Husserl rarely employs the notion of *Horizontbewußtsein* in his early analyses of horizon-intentionality, there is a good sense in which one can speak of a static determination of horizon-consciousness. After all, Husserl early on identifies intentionality as consciousness, taken in its pregnant sense (Hua III/1, 187), while also emphasizing that the horizon is the fundamental feature of intentionality (Hua III/1, 187–191).

As we have seen, the notion of the horizon emerges in *Ideas I* before any explicit inquiry into the *epochē* and the reductions. As seen from the phenomenological standpoint, the natural consciousness at this early stage is still unaware that this horizonal extension, which gets to be equated with the natural world, is in truth the "extension" of consciousness, i.e., that what consciousness confronts here is *lived-space* and *lived-time*. This is the lesson consciousness learns as it performs the transition from the natural to the phenomenological attitudes, i.e., once it performs the *epochē* and the reduction. Due to these methodological procedures, what earlier appeared as a characteristic of objects that are precritically taken to exist "in themselves," now proves to be a feature of consciousness. Yet through these procedures, consciousness itself undergoes significant transformations. It now reveals itself as essentially two-sided, i.e., as embracing both the level of appearances (Hua III/1, §35, §83) and the level of lived-experience (Hua III/1, §82).

This twofold manifestation of the horizon calls for a phenomenological clarification. More particularly, it calls for an explanation of *what kind of structure consciousness itself must possess* so that the horizonality of experience could manifest itself within the field of consciousness as two-sided. In *Ideas I*, the description of consciousness as horizonal provides the phenomenological basis for the interpretation of consciousness in terms of its *noetic* and *noematic* components. That each and every appearance is given within the framework of background appearances and that each and every lived-experience emerges within the stream of *Erlebnisse* is not a consequence that stems from the account of consciousness as noetic/noematic. On the contrary, it is the recognition of the horizonal framework of consciousness itself that underlies the structural account of consciousness as correlational. Otherwise put, *the incentive to speak of noetic and noematic phases of consciousness lies within horizonality itself.*

This is of paramount significance. Such a state of affairs allows one to suggest that, as Husserl's description of the two-sided manifestation of the horizon gives rise to the static notion of horizon-consciousness, this static notion, as an *interpretation* of this two-sidedness in terms of *noesis* and *noema*, nonetheless leaves out a number of dimensions of horizonality and thus leaves the space open for the subsequent emergence of the genetic notion of the horizon.

Horizontbewußtsein as a static notion can be characterized in terms of three layers of sense, each of which presupposes the subsequent one as its phenomenological basis. (1) We begin with pregiven objectivities, be they real or ideal, and approach them in terms of how they show themselves within the context of their background-appearances, which we identify as *noematic* horizons of sense. (2) We further realize that these noematic horizons are themselves built upon more rudimentary *noetic* horizons. (3) Taken in its fullness, noetic intentionality is nothing other than the stream of consciousness itself, which retains its unity due to the constant presence of the *ego*, conceived as an irreducible transcendence within immanence.

To use an example, (1) the flock of geese flying in front of my window can be said to appear in the midst of other objects only if the flock retains its identity throughout its numerous appearances. These appearances reveal the object in question as continuously different, yet still identical, in the midst of the constantly changing background configurations. (2) Yet not only the flock of geese as an appearance has its continuously shifting background of sense; my *seeing* of this object is an *Erlebnis* which has its own unique horizons in that my seeing *intends this very object* in continuously shifting perspectives. (3) This seeing itself is an *Erlebnis* that is given in midst of all other past and future *Erlebnisse*, which intend a variety of objectivities, and only within this indefinite unity can the *Erlebnis* in question manifest itself in its given configuration.

Within such a framework, to thematize the horizon as *Horizontbewußtsein* is to argue that (1) each and every conscious intention co-intends a whole system of intentions, due to which (2) each and every appearance is itself given within a system of appearances. This does not mean that consciousness gives rise to two streams of objectification, one of which would pertain to the totality of intentions, while the other one would give rise to the totality of appearances. To reiterate, here we are dealing with *consciousness as correlational*, as *irreducibly two-sided*. The distinction in question rather suggests that consciousness can be in possession of appearances of objectivities in virtue of its horizonal framework. The horizon thereby simultaneously determines consciousness as a "consciousness-of" and each and every appearance as an "appearance-of." This means that only as horizonal can consciousness be in possession of objects as something meant; and only so can it be in the presence of objects, be they real or ideal. Such, in short, is the static notion, which accounts for the need to thematize the horizon as the horizon of consciousness.

This notion is not without problems. Let us ask: what is the basis of the claim that consciousness in each and every act co-intends a whole system of intentions? Husserl's answer points in the direction of the noetic animation of hyletic data.[16] Yet such an explanation, as my subsequent remarks will suggest, severely restricts our understanding of the horizon.

As Husserl has it, "that which is 'transcendentally constituted' 'on the ground of' the material (*stofflich*) experiences 'by' the noetic functions is, to be sure, something 'given;' and in pure intuition we faithfully describe the experience and its noematic object intended to, it is something *evidently* given; but it belongs to experience in a sense entirely different from the sense in which the really inherent and therefore proper constituents belong to experience" (*Ideas I*, 239).[17] It is by far not clear if such a constitutive framework can be defended on phenomenological grounds at all. First, it is by far not clear if Husserl's account of how noeses animate the hyletic elements of consciousness can be, as Husserl has it, "in reiner Intuition treulich beschrieben" ("faithfully described in pure intuition"). What is actually given in intuition is the *perceived* object and the *perceiving* of this object, but not the manifold acts animating sense data. Secondly, and more significantly, Husserl's account of how each and every intention is composed of numerous noetic phases, each of which animates hyletic data, breaks apart the unity of the act by dissolving it into manifold phases. One seems to be invited to search for the ground of the act's unity within the hyletic level of consciousness, yet according to Husserl's own admission, here no unity is to be found, since unity arises from the animating acts of consciousness. However, on the noetic level, Husserl does not explain how exactly noeses generate the unity of the act, i.e., how exactly an act unifies its own phases. And just as it is unclear how the noesis can have its own phases, so also, by extension, it remains unclear how the noema can have its own inner-horizon.

Thirdly, this very same problem reemerges at the level of noetic-noematic correlation, thus forcing one to wonder if *Ideas I* only proclaims, or indeed accounts

---

[16] See Hua III/1, §85.

[17] "Das 'auf Grund' der stofflichen Erlebnisse 'durch' die noetischen Funktionen '*transzendental Konstituierte*' ist zwar ein 'Gegebenes' und, wenn wir in reiner Intuition das Erlebnis und sein noematisch Bewußtes treulich beschreiben, ein *evident* Gegebenes; aber es gehört eben in einem völlig anderen Sinne dem Erlebnis an, als die reellen und somit eigentlichen Konstituentien desselben." (Hua III/1 228)

for the essential characteristic of the horizon, i.e., for how each and every act co-intends a whole system of intentions. What phenomenological assurance can one have that to each and every noematic phase there indeed corresponds a noetic phase? Along with Jitendra Mohanty, let us ask: does this assumption not bring in a constancy hypothesis and thus the objective prejudice?[18] But if this correlation remains unexplained, then one can have no assurance that each and every act entails a reference through which it co-intends a whole system of other acts. And just as one loses the basis to speak of the co-intention of a system of acts, so also, by extension, one loses the ground on which to account for the noema's outer-horizon.

Thus the noetic/noematic account of consciousness leaves one at a loss as to what structure consciousness must possess if it is to be called horizonal. There is, moreover, yet another reason why this notion of *Horizontbewußtsein* is deficient, and the reason for this deficiency is at the same time the reason why this notion of horizon-consciousness is a *static* notion. This notion provides a *structural account* of the continuous presence of the horizons in each and every conscious act; yet this notion remains silent when one asks why with each and every act of consciousness, the horizons, their continuous presence notwithstanding, undergo continuous change. The static notion of the horizon suppresses the *fluidity* that lies at the heart of the horizon; it does not show why the configurations of sense that the horizons bring forth are never only formal configurations, but that the projections of sense, which the horizons initiate, are always already filled with continuously changing content. Not only are the horizons of sense, which engulf different objectivities, different from each other; each and every objectivity reveals itself as continuously different due to the constantly changing horizons that embrace it. Yet how can the content of objectivities be always different? In *Ideas I*, Husserl provides us with the following answer: "the object, tree, *can only* appear *at all* in a perception as *Objectively* determined in the mode in which it does appear in the perception if the hyletic moments are just those and no others. This therefore implies that any changes of the hyletic content of the perception ... must at least result in what appears becoming objectively 'other,' whether in itself or in the orientation in which it is appearing, or the like" (*Ideas I*, 238).[19] Yet such a response, as it simply points to the changes that lie within the hyletic level, is in short supply. It only invites one to redirect our question to the hyletic level: what exactly causes the hyletic data to change? Within the framework of *Ideas I*, this question remains unanswered.

Thus the static notion of horizon-consciousness only partly discloses what is entailed in the horizon-problematic. Nonetheless, one should not overlook that almost each and every *exemplification* of the horizon in *Ideas I* entails the dimension of persistent change. Besides being exemplified, this crucial dimension of

---

[18] See Mohanty (1972), pp. 96–107.

[19] "[...] der Gegenstand Baum [kann] in einer Wahrnehmung *überhaupt* als *objektiv* so bestimmter, wie er in ihr erscheint, *nur dann* erscheinen, wenn die hyletischen Momente [...] gerade die sind und keine anderen. Darin liegt also, daß jede Änderung des hyletischen Gehaltes der Wahrnehmung [...] zum mindesten den Erfolg haben muß, daß das Erscheinende zu einem objektiv "anderen" wird." (Hua III/1, 227)

horizonality has not completely escaped thematization either. In §44, for instance, in the context of discussing the horizon as co-givenness that encircles appearances as "more or less vague indeterminacy" (Hua III/1, 91), Husserl writes:

> This indeterminacy necessarily signifies a *determinableness which has a rigorously prescribed style*. It *points ahead* to possible perceptual multiplicities which, margining continuously into one another, join together to make up the unity of one perception in which the continuously enduring physical thing is always showing some new 'sides' in a new series of adumbrations. (*Ideas I*, 94)[20]

Due to this continuous change that qualifies the appearance of transcendent objectivities, Husserl obtains the ground on which he can distinguish between, as the title of §44 of *Ideas I* declares, the merely phenomenal being of something transcendent and the absolute being of something immanent—a distinction, which I have already addressed in Chap. 5 and to which I will still return in Sect. 7.2. Yet the manner in which the horizon-consciousness projects a continuously different configuration of sense remains unexplored. One can therefore claim that the horizon in its phenomenality entails dimensions of sense that lie beyond Husserl's early thematization of the horizon. Let us see how the thematization of the so-far unclarified dimensions of the horizon generates a decidedly new notion of horizon-consciousness.

6.4 The Genetic Notion of Horizon-Consciousness

On what grounds can consciousness be said to be "consciousness-of," i.e., how can consciousness be in possession of objectivities *as something meant*? To say, as I have suggested in the previous section, that this is an achievement of horizon-consciousness is, although justified, not sufficient, for this response does not clarify either the manner in which horizon-intentionality *obtains*, or how it *entails*, the configurations of sense which embrace appearing objectivities. With this issue in mind, let us turn to Husserl's direct analysis of horizon-consciousness in §33 of *Experience and Judgment* as representative of the manner in which horizon-consciousness is thematized in genetic phenomenology.

In this section—one of the few in the whole corpus of Husserl's works in which horizon-consciousness is not just operative, but directly thematized—Husserl speaks of horizon-consciousness as "the horizon of typical pre-acquaintance in which every object is pregiven" (EJ, 150).[21] It is hard to overestimate the significance of this brief characterization: by disclosing what this remark entails, one can derive the main features of the genetic dimension of horizon-consciousness.

---

[20] "Die Unbestimmtheit bedeutet ja notwendig *Bestimmbarkeit eines fest vorgeschriebenen Stils. Sie deutet vor* auf mögliche Wahrnehmungsmannigfaltigkeiten, die, kontinuierlich ineinander übergehend, sich zur Einheit einer Wahrnehmung zusammenschließen, in welcher das kontinuierlich dauernde Ding in immer neuen Abschattungsreihen immer wieder neue "Seiten" zeigt." (Hua III/1, 91)

[21] "[...] der Horizont typischer Vorbekanntheit, in dem jeder Gegenstand vorgegeben ist." (EU, 172)

To characterize horizon-consciousness as "a horizon of typical pre-acquaintance" is to suggest that consciousness immediate *recognizes* each and every appearance as an appearance of a *particular type* of objectivity, due to which an appearance is immediately *apperceived* within an "inductive" or anticipatory framework (EU, 28). The awakening of typical pre-acquaintance is the awakening of the horizon, i.e., of the other possible manners of givenness, which the anticipatory framework prescribes to the appearing phenomenon. Husserl goes on to suggest that

> that which is for us an object self-given in a simple intuition, such that it can be apprehended in its internal as well as in its relative characteristics, rests, therefore, not only on what is intuitive and self-given itself and capable of being self-given intuitively as the object's surrounding field of intuition. It rests also on all the relations—which for the most part remain undisclosed—to what has been once given and which can possibly be representified, indeed, possibly on all the relations to the objectivities… (EJ, 150).[22]

However, the recognition that each and every appearance awakens horizons of consciousness as the horizons of typical pre-acquaintance should not be taken to mean that the emergence of the "undisclosed relations to what was once given" is fixed from the start. Rather, the horizons are awakened as horizons of *indeterminate determinability*. This means that appearance in its duration comprises not only permanent fulfillment, but also elements of frustration, which modify the projected configuration of sense that accompanies the appearing objectivity. Thus the horizon of typical pre-acquaintance is *defining*, yet not *definite*, *determining*, although not *determinate*. The incessant modifications that the horizon is capable of undergoing indicates that the horizon, besides ruling over appearing phenomena, is in its own turn governed by them. We are facing a *codetermination* of appearances and horizon-consciousness not only because the horizons are motivated by appearances in their *emergence*, but also because appearances in their *duration* continue to compel horizon-consciousness to obtain and exert new kinds of anticipatory projections. Thus the horizons determine appearances as appearances of a particular objectivity; yet it is appearances themselves that motivate the horizons to schematize them in a called-for manner. For instance, so as to perceive a deck of cards on the table as cards and not as boxes of matches, or so as to see the birds flying over the river as birds and not as autumn leaves, consciousness must be guided by the awakened horizons of anticipation. Yet it is the appearing phenomenon in its duration that motivates consciousness to hold on to the projected configuration of sense; and when appearances frustrate the projected configuration, the horizons undergo a modification: what earlier showed itself as a deck of cards now proves to be a box of matches; or what I see now is no longer a flock of geese, but autumn leaves, which the wind has torn from the trees. Due to such a codetermination, the horizons reveal themselves not only in terms of their formal structure, but also as concrete

[22] "Das, was ein schlicht anschaulich selbstgegebener Gegenstand für uns ist, als was er in seinen inneren und relativen Beschaffenheiten erfaßt werden kann, das beruht also nicht nur auf dem, was anschaulich an ihm selbst und als sein anschauliches Umfeld selbstgegeben und selbst gebbar ist, sondern auch auf all den zumeist unenthüllt bleibenden Verhältnissen zu einmal gegeben Gewesenem, [nur] eventuell wieder zu Vergegenwärtigendem[…]." (EU, 172–73)

projections of sense, which call for the object to manifest itself in an anticipated way. Moreover, once taken in their full concreteness, the horizons reveal themselves as capable of continuous modification, and thus the plain fact that the horizons are continuously changing no longer brings us to a loss, as it did while we were operating with the static notion of horizon-consciousness.

To say that the horizons of sense that envelope particular objectivities are not fixed from the start is to suggest that the course of experience can always modify the horizonal framework of consciousness. Such continuous modifications are possible due to the fact that the accomplishment of appearances is not exhausted by the givenness of particular objectivities. Besides being necessary conditions for the experience of objectivity as something meant, appearances also generate new *types* of objectivities, thereby enabling consciousness to project a new type of an anticipatory horizon when confronted with new appearances. We thereby see *plasticity* as a central characteristic of horizon-consciousness: the latter just as much determines appearances as "appearances-of" as it itself is transformed by appearances. On the one hand, each and every experience is experienced as far as it is already anticipated by horizon-consciousness; on the other hand, the new experience merges with the horizon itself, thereby transforming the horizon's anticipatory schema. The indeterminacy of the horizon thereby shows itself as inherently productive.

Yet does this codetermination of appearances and horizon-consciousness not land us in a vicious circle? Do we not find ourselves torn between two incompatible tendencies, viz., by the need, on the one hand, to recognize the horizonal framework as necessarily *antecedent* in relation to each and every appearance, and, on the other hand, by the need to admit that horizon-consciousness itself *emerges from* those sedimentations of sense that stem from appearances? Just as an appearance cannot be an appearance before being "schematized," so a horizon cannot be completely empty, i.e., a horizon without anything appearing "within" it.

§33 of *Experience and Judgment* provides a helpful hint regarding how this difficulty can be resolved. In this section Husserl suggests that the horizon of typical pre-acquaintance "has its ground in the passive associative relations of likeness and similarity, in the 'obscure' recollections of the similar" (EJ, 150).[23] This brief reference to the syntheses of homogeneity and heterogeneity is significant in that it places a demand to further deepen the notion of horizon-consciousness. The horizon is a horizon of consciousness not only because each appearance leaves an imprint upon consciousness thereby generating particular horizons of objectivity, but also because consciousness needs to have provided an associative (and, even more fundamentally, a temporal) horizon within which an appearance could emerge as an appearance. Thus the syntheses of *homogeneity* and *heterogeneity*—the fundamental associative syntheses—are further-reaching dimensions of horizon-consciousness.

[23] "Aber nicht nur das originaliter wahrnehmbar im gegenständlichen Hintergrund Mitgegebene bietet Anlass zu beziehender Betrachtung und Gewinnung von relativen Bestimmungen, sondern auch der Horizont typischer Vorbekanntheit hat ja ihren Grund in passiven, assoziativen Gleichheits- und Ähnlichkeitsbeziehungen, in 'dunklen' Erinnerungen an Ähnliches." (EU, 172)

Here we need to contend ourselves with a brief indication that in his account of these syntheses, Husserl shows how an appearance of a prominence (*Abgehobenheit*) as a rudimentary appearance necessarily emerges within a horizon of sense which these syntheses delineate (EU, §16).

Yet is the theme of "typical pre-acquaintance" robust enough to generate a notion of horizon-consciousness different from the one operative in *Ideas I*? I would suggest that such indeed is the case. The recognition of the necessary emergence of the horizons of typical pre-acquaintance determines consciousness in a decidedly new fashion. The awakening of typical pre-acquaintance unfolds consciousness not only as a stream which *despite its streaming* retains an identical noetic/noematic structure, but rather, and more importantly, as a *field which, due to its streaming, is composed of sedimented accomplishments*. Horizons are *horizons of consciousness* because the configurations of sense they project have their origins in the sedimentations of consciousness. Horizons are projections of sense which consciousness itself brings forth by way of awakening what is sedimented through the past accomplishments of consciousness.

In virtue of the horizons of typical pre-acquaintance, we obtain a deeper understanding of what it means to characterize consciousness as a stream. The latter is not to be understood merely in terms of its formal structure, but rather as a field that condenses the sedimentations of sense, i.e. a field into which each and every *Erlebnis* "streams in" and leaves its permanent imprint. It is a stream that continuously generates new *habitualities*, which in their own turn compel consciousness to apperceive appearances as appearances of determinate objectivities.

Through the sedimentations of sense that generate habitualities, we obtain a decidedly novel understanding of what it means to characterize the horizon as horizon-consciousness. This characterization suggests that the horizon is a configuration of projected sense that accompanies each and every actual appearance, and that this configuration emerges from consciousness as a store of lasting possessions. We thereby also obtain a clearer sense why consciousness from the start can be characterized as "consciousness-of." By immediately awakening the horizons of typical pre-acquaintance, each and every appearance shows itself as necessarily something more than it itself is, i.e., as an appearance of objectivity.

My characterization of the genetic notion of horizon-consciousness is not yet complete and the reason for this lack lies in the ambiguity that qualifies the expression "consciousness-of." Not only does this notion indicate an intentional bond that ties consciousness to objectivity; it just as much ties consciousness back to subjectivity itself: *consciousness is consciousness of the ego*.[24] To characterize consciousness as *my* consciousness does not only mean to privilege one methodological standpoint over others; it also means to address consciousness the way it gives itself, i.e., address it in its concreteness. Only through a methodological abstraction can we cut off the bond that ties consciousness to subjectivity.[25] The privileging of the

[24] See Zahavi (2005), particularly the analysis in Chap. 5.

[25] "Leben ist zunächst egologisches Leben und als mein, des sich Besinnenden, Leben[…]. Dabei ist es ichzentriertes Leben der Passivität, Affektivität und Aktivität." (Mat. VIII, 42)

"Ich-Rede" in Husserl's phenomenology[26] should not be understood as a consequence that stems from an arbitrary methodological decision; it rather derives from the recognition that consciousness can only remain abstract if treated separately from subjectivity.

The awakening of typical pre-acquaintance is possible only as far as consciousness is stamped by "mineness:" Otherwise put, the sedimentations and habitualities of consciousness are possible only so far as they at the same time are *my* sedimentations and *my* habitualities. The "mineness" in question should not be understood as an enigmatic substrate which, with the help of some kind of *Deus ex machina*, comes into a relation with the stream of consciousness.[27] Rather, as thematized in Husserl's static and genetic inquiries, *"mineness" is the self-centering of consciousness itself.* As Husserl puts it in a text that belongs to the C-group of manuscripts, "The I is the 'subject' of consciousness. The 'subject' is only a word for the centering, which has all life as my-life, and thereby can live through everything that it lives through, and be conscious of everything it is conscious of" (Mat VIII, 35).[28] Without this dimension of self-centering, the themes of habitualities and sedimentations would lose their sense, for these "lasting possessions" are possible only if consciousness has a dimension of *self-centering*. We thus see that *horizon-consciousness, once taken in its concreteness, shows itself as the horizon of subjectivity.*

Yet the inseparability of consciousness and of the I should not overshadow the difference between them: to qualify the I as the "centering point" of consciousness is already to suggest that the stream of consciousness resists being fully assimilated by the I. So as to make sense of this difference and this resistance, let us recall the foregoing analysis of the selective and discriminative nature of experience in Sect. 5.2. Following Husserl's claim "the seizing-upon is a singling out and seizing" (*Ideas I*, 70),[29] I argued that appearances need not be thematic in order to be given; I further suggested that the originary givenness of appearances is non-thematic as well as non-explicit. The distinction between thematic and non-thematic appearances provides a clue as to how the difference between consciousness and the I should be conceived. Husserl's recognition of the different types of appearances intimates that appearances can be given within *the stream of consciousness* without being recognized as thematic objects of *the I*. The horizon thereby manifests itself in virtue of the tension between consciousness and the I, when the I is conceived as the self-centering of consciousness. We thereby face a tension between consciousness and its "mineness," and it is this very tension that allows us to identify the structure of background/foreground as the central structure

[26] See, for instance, Hua VIII, 59.

[27] We find Husserl's critique of such a notion of the ego in the Fifth of the *Logical Investigations*. Also, in its main lines, this notion of the ego is Husserl's central target of criticism of Descartes that we find in the First *Cartesian Meditation*.

[28] "Das Ich ist 'Subjekt' des Bewußtseins. 'Subjekt' ist dabei nur ein anderes Wort für die Zentrierung, die alles Leben als Ichleben—und somit lebend etwas zu erleben, etwas bewußt zu haben—hat." (Mat VIII, 35)

[29] "Das Erfassen ist ein Herausfassen." (Hua III/1, 71)

of horizon-consciousness.[30] Even more: *the tension between consciousness and mineness is what accounts for the possibility of the background/foreground structure of appearances*: in virtue of the ego's centering on a particular mode of givenness, consciousness is in possession of a *thematic* objectivity; in virtue of the fact that the ego cannot appropriate all that is given to consciousness, the latter is also in possession of *non-thematic* modes of givenness.

Such a difference between consciousness and the ego invites one to consider an intriguing possibility. In virtue of the tension between consciousness and the ego, is one not in full right to speak of "horizonless experiences" and "horizonless appearances?" For good reasons, these notions are not cases of *contradictio in adjecto*: if the ego is the *centering* of consciousness, then horizonless experience and horizonless appearance are rendered possible when appearances are intended in, and experiences undergone by, a *decentered consciousness.* Such seems to be at times the givenness of appearances in dreams or daydreaming. And arguably, if one were capable of winding back the clock and slowing down the ticking of its beats, one would find oneself facing a great number of such decentered experiences. To be sure, by making them thematic, one would immediately misapprehend them, since for consciousness to zero-in on an appearance means nothing other than to place it within a horizon of sense—a horizon that did not embrace it initially. Yet after such an experiment, one would be in the position to recognize that horizons can only follow in the footsteps of the prior givenness of appearances, viz., the givenness of appearances to consciousness, which resists any kind of assimilation.

And yet, as I have already suggested in the introduction to this study, the experience subjectivity undergoes in dreams and daydreaming is not horizonless. Rather, the resistance of consciousness to be assimilated by the I indicates subjectivity's capacity to inhabit a number of different horizons of sense. The foregoing analysis suggests that this capacity derives from the fact that consciousness *per se* does not have its own horizons; the horizons belong to consciousness only insofar as *consciousness itself is consciousness of the ego*. It is the ego that projects horizons of sense that embrace appearances, which in their own turn can be given to consciousness even in the absence of the above-mentioned assimilation. To put the matter briefly, insofar as appearances are given to consciousness and to the ego, appearances are horizonal; insofar as they are given to decentered consciousness, they are horizonless.

All of this might sound enigmatic, especially in the face of Husserl's notion of the ego that he employs in the confines of static phenomenology. Indeed, our understanding of the tension between consciousness and the ego remains almost

[30] Let it be remarked in passing that, as we have already seen in Chap. 5 and as the discussion of the world as the *wherefrom* of experience in the third part of this work will show again, the structure in question is not exhaustive of Husserl's notion of the horizon. There is a sense in which a background without a foreground, a halo without any kind of intuitive core, is a figure of horizon-intentionality. Even more, this figure is the most rudimentary manifestation of the horizon in Husserl's phenomenology.

inscrutable for as long as we do not take into account the enrichment of sense that the notion of the I undergoes in Husserl's phenomenology. When the notions of the ego, the I, or subjectivity are employed in genetic phenomenology, they are significantly different from the notion of the "pure ego" as encountered in *Ideas I*. So as to specify the difference in question, let me turn to Husserl's thematization of three different notions of the ego in *Cartesian Meditations*.

6.5 The Ego in Its Static and Genetic Determinations: The Emergence of Transcendental Subjectivity

No other of Husserl's works reveals the transformations that the notion of the ego undergoes more lucidly than *Cartesian Meditations*. In §30–§33 of this programmatic text, Husserl progressively deepens our phenomenological understanding of the ego by providing this notion with three determinations. The first of them is essentially identical with the one already encountered in *Ideas I*. *The ego is said to be the identical pole of Erlebnisse*.[31] Just as the object polarizes appearances by way of stamping them as appearances of this particular object, so the ego also polarizes lived-experiences by way of stamping them as experiences of this particular ego. Just as the object is transcendent to all of its *cogitata* while at the same time being inseparable from them, so the ego too is transcendent to all of its *cogitationes*, even though they are indivisible from it.

One can characterize the motivation that underlies the disclosure of this notion of the ego as a refusal to provide the "I" with any qualities or content so as to retain its non-objective "nature" and thus, in the words of Shigeru Taguchi, to obtain the evidence of the ego as the "evidence of closeness" ("Evidenz der Nähe"). As Taguchi has it, "That extreme 'simplicity' and lack of content of the pure I indicate that nothing objective can be attested to this ego. The evidence for it can be neither reinforced nor corrected, since it does not pass through any kind of variable content. It can only persist in the simple 'I am,' 'I live,' that is, 'I function'" (Taguchi 2006, 72–73).[32] And yet, I would like to emphasize that even though the notion of the ego as the identical pole of *Erlebnisse* aims to capture the ego's concreteness, nevertheless, the enrichment of this notion—an enrichment that takes on the form of an

[31] Consider also how this notion is spelled out in Hua XIV: "Ein Ich hat keine generelle sachhaltige Eigenart; es ist an solcher gänzlich leer. Es ist bloß ego des cogito, das den ganzen Gehalt abgibt, und bezogen auf einen Erlebnisstrom, dem gegenüber es auch unselbständig ist, wie auch umgekehrt." (Hua XIV, 23) In Hua IV, Husserl describes the pure ego in the following manner: "Als reines Ich birgt es keine verborgenen inneren Reichtümer, es ist absolut einfach, liegt absolut zutage." (Hua IV, 105)

[32] "Jene extreme 'Einfachheit' und Inhaltslosigkeit des reinen Ich besagt, daß von ihm eigentlich nichts objektiv zu bestätigen ist. Diese Evidenz kann weder bekräftigt noch korrigiert werden, da sie über keinen variierbaren Inhalt verfügt. Sie kann nur im einfachen Selbstvollzug des 'Ich bin, ich lebe' bzw. 'Ich fungiere' bestehen[...]." (Taguchi 2006, 72–73)

abandonment of some of this notion's central characteristics—is motivated by the recognition that the notion under consideration is by far not concrete.

Indeed, as the analogy between the ego and the intentional object in its main features reasserts the notion of the ego as thematized in *Ideas I*, it simultaneously forces one to deepen its initial portrayal and to abandon some of its central characteristics. More precisely, it compels one to give up the supposition that the ego is an *empty* pole of identity. Just as the object is not an empty pole, since its sense is directly affected by its different modes of appearance, so the ego too is not empty, since it is also dependent upon the stream of *Erlebnisse*. So as to thematize this dependence, Husserl provides the ego with its second determination and speaks of it as *the substrate of habitualities*. One should not overlook that, as Husserl remarks repeatedly, by obtaining this second determination of the ego, we step into the threshold of genetic phenomenology.[33]

The ego is transcendent and simultaneously inseparable from its acts not only because it is the centering source (here Husserl speaks of the "zentrierende Ich") from which each act emanates. Rather, acts leave an imprint upon the ego, due to which the ego obtains personal features. *Erlebnisse* "stream in," yet there is a sense in which they do not "stream away": "I decide for the first time in favor of a being and a being-thus; the fleeting act passes, but from now on, *I am abidingly the Ego who is thus and so decided*" (CM, 66).[34] My resolutions and convictions remain truly mine until I willfully abandon them. Yet even their abandonment, just as much as the decisions due to which I obtain them, provide the ego with a lasting *habitus*. In virtue of these decisions and their abandonments, the ego undergoes modifications, due to which, and in spite of which, it obtains an abiding style and a personal character.

Let us ask: can consciousness *per se* have its own sedimentations and its own habitualities? We obtain an answer from Husserl's analysis of the habitualities of the ego. Of significance is the following claim: "the persisting, the temporal enduring, of such determining properties of the Ego, or the peculiar change that the Ego undergoes in respect of them, manifestly is not a continuous filling of immanent time with lived-experiences—just as the abiding Ego himself, as the pole of abiding Ego-properties, is not a lived-experience or a continuity of lived-experiences, even though, with such habitual determining properties, he is indeed related back to the stream of lived-experiences" (CM, 67).[35] The attempt to subscribe habitualities to the stream

[33] At the very outset of his analysis of the ego as the substrate of habitualities, Husserl speaks of the *Gesetzmäßigkeit der "transzendentalen Genesis*." (Hua I, 100) At the very offset of the analysis of all three determinations of the ego, he remarks: "Mit der Lehre vom Ich als Pol seiner Akte und als Substrat von Habitualitäten haben wir schon, und in einem bedeutsamen Punkt, die Problematik der phänomenologischen Genesis berührt, und damit die Stufe der genetischen Phänomenologie." (Hua I, 103)

[34] "Ich entschließe mich—das Akterlebnis verströmt, aber der Entschluß verharrt [...] Ich bin hinfort der so Entschlossene." (Hua I, 101)

[35] "Das Verharren, das zeitliche Dauern solcher ichlichen Bestimmtheiten und das ihnen eigentümliche *Sich-verändern* besagt offenbar keine kontinuierliche Füllung der immanenten Zeit mit Erlebnissen, wie denn das bleibende Ich selbst als Pol bleibender Ich-Bestimmtheiten kein Erlebnis und keine Erlebniskontinuität ist, obschon doch wesensmäßig mit solchen habituellen Bestimmtheiten zurückbezogen auf den Erlebnisstrom." (Hua I, 101)

of consciousness cannot find phenomenological justification because the stream of consciousness is a stream of *Erlebnisse*, and therefore, if habitualities belonged to consciousness, they would have to be *Erlebnisse* themselves. Yet such, clearly, cannot be the case: habitualities do not emerge into and recede from the field of consciousness the way *Erlebnisse* do; they rather stamp *Erlebnisse* with a particular sense-determination in a manner which indicates their simultaneous inseparability and transcendence from the stream of experience (*Erlebnisstrom*). *Habitualities are "within" consciousness, yet not "of" consciousness.* Their unique relation to consciousness is nothing other than the unique relation of consciousness to the ego.

But if so, then the horizonal framework, as the last section already suggested, in truth is also a framework that belongs to the ego. With the notion of the ego as the substrate of habitualities, we obtain a phenomenologically justifiable way in which consciousness can be said to have its own horizons: the horizon is the horizon-consciousness, but only insofar as it is a *horizon-consciousness-of-the-ego*.

Just as the first determination of the ego, so the second one also is not yet fully concrete. The ego is the identical pole and the substrate of habitualities not only, and not primarily, in regard to the stream of *Erlebnisse.* Yet it is precisely this relation that exhausts the framework within which the ego obtains its first two determinations. Hence the need of a third notion which would thematize the ego in terms of the very life it lives, i.e., in terms of the relation of the ego to *cogitata*, to intentional objectivities, and to the surrounding world (*Umwelt*) in which it always finds itself. This third notion, which Husserl, borrowing a central concept from Leibniz, calls the *monad*, envelopes "the whole actual and potential life of consciousness" (Hua I, 102). To uncover what this life entails is nothing less than to unfold the ego's intentional life as it manifests itself in the process of constitution. Such a thematization does not abandon the first two determinations of the ego, but rather explicates them in their *concreteness*, by showing how the ego retains its transcendence and how it obtains its habitualities in the process of transforming appearances into full-fledged objectivities. One could say that the ego as a monad is a *depsychologized* version of the notion of the ego as the substrate of habitualities. The significance of this last notion is of paramount significance for phenomenology. As Husserl has it, "it is clear that the problem of *explicating this monadic ego phenomenologically* must include *all constitutional problems without exception*" (CM, 68).[36]

The second notion of the ego is not yet fully concrete because it does not spell out how the habitualities of consciousness affect the sense of objectivities. The third notion deepens the second one by revealing the role that habitualities play within the constitutive framework. Here we are facing an analogical relation between horizon-consciousness and objectivities to the one we encountered already in *Ideas I.* Yet there is a crucial difference: horizon-consciousness in its genetic variant is shown to "schematize" objectivities not only in terms of the noetic-noematic

[36] "Da das monadisch konkrete ego das gesamte wirkliche und potentielle Bewußtseinsleben mit befaßt, so ist es klar, daß das Problem der phänomenologischen Auslegung dieses monadischen ego (das Problem seiner Konstitution für sich selbst) *alle konstitutiven Probleme überhaupt* in sich befassen muß." (Hua I, 102–103)

structures that underlie them but, more importantly, in terms of the *content* that belongs to concrete objectivities. Due to such a deepening, Husserl insists that "the phenomenology of this *self-constitution* coincides with *phenomenology as a whole*" (CM, 68).[37]

Admittedly, Husserl's discussion of the three determinations of the ego in *Cartesian Meditations* is of a cursory nature. My task here, however, is not that of providing a full account of what this notion entails in Husserl's phenomenology. Rather, what I want to emphasize and to pursue is that in contrast to the notion of the pure ego of *Ideas I, which has no horizons of its own*, the notion of the full subjectivity as thematized in the *Meditations* is a notion, which entails a unique horizonality of subjectivity. *That subjectivity itself possesses its own unique horizons can be considered one of the most fundamental accomplishments of Husserl's genetic analysis of horizon-intentionality.*[38] Due to this horizonality, it will prove possible to broach anew the question of the origins of the horizons in Husserl's phenomenology in a manner decidedly different from the historical way I addressed this question in the first part of this work.

That the ego has its own horizons is already entailed in the notion of the ego as a substrate of habitualities. What this notion encompasses was already addressed in the discussion of the genetic notion of horizon-consciousness. Yet neither the foregoing discussion, nor the present analysis of the ego as the substrate of habitualities, is sufficient to generate an account of horizon-consciousness in all its concreteness. Both remain to a degree abstract in that they do not explain how one is to understand the relation between the ego's horizonality and those horizonal structures thematized in the first part of this work.

Emphasizing that the fully concrete notion of the "I" contains the whole actual and potential life of consciousness in that it progressively absorbs new habitualities (Hua I, 102), Husserl points in the direction of a notion of the ego which entails more rudimentary dimensions of horizon-intentionality than the ones disclosed in *Ideas I*. That this horizonality is more rudimentary, one can already glimpse from the fact that the horizons that envelop particular objectivities (and the surrounding world within which they emerge) can manifest themselves only on the basis of more basic accomplishments of consciousness. Given that consciousness, insofar as it belongs to the ego, itself has its own horizons, one is compelled to contend that the horizons of consciousness are more rudimentary in that they "unfold the space" within which the already analyzed horizon-structures can manifest themselves. Admittedly, Husserl's all-too-brief account of the third notion of the ego in *Cartesian Meditations* does not disclose the main features of this rudimentary horizonality. Even worse: this analysis does not clarify how this so-far anonymous horizonality can be uncovered. With the last issue in mind, let me turn to Husserl's discussion of the reduction and of the full subjectivity in *First Philosophy II*.

[37] "In weiterer Folge ergibt sich die Deckung der Phänomenologie dieser Selbstkonstitution mit der Phänomenologie überhaupt." (Hua I, 103)

[38] The other central accomplishment, as Part III of this investigation will maintain, lies in the discovery of the world-horizon.

References

Gadamer, Hans-Georg. 1989. *Truth and method.* Trans. Joel Weinsheimer. New York: Continuum.

Gadamer, Hans-Georg. 1990. *Wahrheit und Methode. Grundzüge einer philosophischen Hermeneutik.* Tübingen: J.C.B. Mohr (Paul Siebeck).

Heidegger, Martin. 2006. *Sein und Zeit.* Tübingen: Max Niemeyer Verlag.

Husserl, Edmund. 1952. In *Ideen zu einer reinen Phänomenologie und phänomenologischen Philosophie. Zweites Buch: Phänomenologische Untersuchungen zur Konstitution*, ed. Marly Biemel. The Hague: Martinus Nijhoff.

Husserl, Edmund. 1959. In *Erste Philosophie. Zweiter Teil: Theorie der phänomenologischen Reduktion*, ed. Rudolf Böhm. The Hague: Martinus Nijhoff.

Husserl, Edmund. 1960. *Cartesian meditations.* Trans. Dorion Cairns. Dordrecht: Kluwer Academic Publishers.

Husserl, Edmund. 1966. In *Analysen zur passiven Synthesis: aus Vorlesungs- und Forschungsmanuskripten, 1918–1926*, ed. Margot Fleischer. The Hague: Martinus Nijhoff.

Husserl, Edmund. 1969. *Formal and transcendental logic.* Trans. Dorion Cairns. The Hague: Nijhoff.

Husserl, Edmund. 1973a. In *Cartesianische Meditationen und Pariser Vorträge*, ed. Stephan Strasser. The Hague: Martinus Nijhoff.

Husserl, Edmund. 1973b. In *Zur Phänomenologie der Intersubjektivität. Texte aus dem Nachlass. Zweiter Teil. 1921–28*, ed. Iso Kern. The Hague: Martinus Nijhoff.

Husserl, Edmund. 1973c. *Experience and judgment: Investigations in a genealogy of logic.* Trans. James Churchill and Karl Ameriks. Evanston: Northwestern UP.

Husserl, Edmund. 1974. In *Formale und transzendentale Logik. Versuch einer Kritik der logischen Vernunft*, ed. Paul Janssen. The Hague: Martinus Nijhoff.

Husserl, Edmund. 1976. In *Ideen zu einer reinen Phänomenologie und phänomenologischen Philosophie. Erstes Buch: Allgemeine Einführung in die reine Phänomenologie*, ed. Karl Schuhmann. The Hague: Martinus Nijhoff.

Husserl, Edmund. 1983. *Ideas pertaining to a pure phenomenology and to a phenomenological philosophy. First book: General introduction to a pure phenomenology.* Trans. F. Kersten. The Hague: Martinus Nijhoff.

Husserl, Edmund. 1999. In *Erfahrung und Urteil. Untersuchungen zur Genealogie der Logik*, ed. Ludwig Landgrebe. Hamburg: Felix Meiner Verlag.

Husserl, Edmund. 2001. *Analyses concerning passive and active synthesis. Lectures on transcendental logic.* Trans. Anthony J. Steinbock. Dordrecht: Kluwer.

Husserl, Edmund. 2006. In *Späte Texte über Zeitkonstitution (1929–1934). Die C-Manuskripte*, ed. Dieter Lohmar. Dordrecht: Springer.

Larrabee, Mary Jeanne. 1976. Husserl's static and genetic phenomenology. *Man and World* 9(2): 163–174.

Mohanty, Jitendra Nath. 1972. *The concept of intentionality: A critical study.* St. Louis: Warren H. Green Inc.

Steinbock, Anthony. 1998. Husserl's static and genetic phenomenology: Translator's introduction to two essays. *Continental Philosophy Review* 31(2): 127–152.

Taguchi, Shigeru. 2006. *Das Problem des 'Ur-Ich' bei Edmund Husserl. Die Frage nach der selbstverständlichen Nähe des Selbst.* Dordrecht: Springer.

Zahavi, Dan. 2005. *Subjectivity and Selfhood. Investigating the First-Person Perspective.* Cambridge, MA: MIT Press.

Chapter 7
The Reduction as the Disclosure of the Horizons of Transcendental Subjectivity

7.1 The Significance of the Reduction for the Thematization of Horizon-Consciousness

As we have already seen, the fundamental contribution of genetic phenomenology to the horizon-problematic consists in the disclosure of the horizon as the *horizon of subjectivity*. Let us take a step back and ask: why did this notion remain absent from static phenomenology in general and from *Ideas I* in particular? The reasons are by far not accidental. I have already indicated one of them: the explanatory schema operative in static phenomenology reduces the content of acts to the hyletic data and thereby fails to account for the horizonal sedimentations that acts entail. Yet this is by far not the only, and not the fundamental, reason. In this chapter, I will argue that the *reduction*, as thematized in *Ideas I*, blocks access to the emergence and manifestation of the genetic phenomenality of the horizon. The horizon-consciousness as a genetic notion thus calls for a broadening of the phenomenological field, which the methodological considerations first have to establish.

A new way to the reduction will enable us attain such a broadening. This new path, which Husserl spells out in *First Philosophy II*, gives rise to the notions of the *double reduction* and of *intentional implications*—notions that ultimately enable us to broach anew the question of the origins of horizon-consciousness.[1] Yet before such a resolution, this path will lead us to a paradox: we will witness how the horizon-consciousness, as it emerges from the new path to the reduction, turns back

---

[1]It is interesting to note that Husserl has already employed the notions of the double reduction and of intentional implications as early as 1910/1911, in his lecture course *Grundprobleme der Phänomenologie*. However, in these lecture notes, Husserl does not draw a distinction between the different paths to the reduction, as he does in *First Philosophy II*. One could thus say that even though these notions are already present in phenomenology before the publication of *Ideas I*, their full import is not disclosed until a clear distinction between static and genetic phenomenology is drawn.

S. Geniusas, *The Origins of the Horizon in Husserl's Phenomenology*, Contributions to Phenomenology 67, DOI 10.1007/978-94-007-4644-2_7,

upon the path which has brought it to its phenomenality and reveals its mundane presuppositions. These presuppositions in their own turn will signify that the horizons in question qualify not so much the transcendental, but rather psychological consciousness.

Yet this apparent dead-end will be overcome by means of radicalizing the reduction and rendering it universal. As universal, the reduction will reveal two distinct dimensions of sense that are entailed in the notion of the horizon: the latter is both *implicit reference* and *implicit validity*. I have already indicated these dimensions of horizonality in Sect. 2.2. Yet my preliminary account of validity and reference as fundamental features of horizonality suffered from two significant limitations: my account neither illuminated the conceptual framework that renders these dimensions of sense possible, nor did it bring to light the implications that these dimensions entail, implications that hold the promise of disclosing the genetic phenomenality of the horizon. Having accounted for the unity of these two dimensions of horizon-intentionality on the methodical and thematic levels, I will proceed to show how the unified notion of the horizon opens up the possibility of inquiring anew into the origins of the horizon in Husserl's phenomenology.

7.2 The Cartesian Path to the Reduction Suppresses the Phenomenality of the Horizon

Is the genetic notion of horizon-consciousness a phenomenological notion? To claim that such indeed is the case, one must first show how the reduction leads to the emergence of the horizon as a phenomenological theme. Yet this cannot be done by following the so-called "Cartesian Way" to the reduction, which Husserl delineates in *Ideas I*.[2] Following Iso Kern, let us single out four central aspects of the Cartesian path: (1) The inquiry into the reduction is guided by the idea of philosophy as an absolutely grounded science built on an absolutely grounded procedure which calls for the beginning to consist in absolute evidence. (2) The philosopher who begins absolutely must put out of action the belief in the world and all knowledge of the world, be it scientific or prescientific. (3) The valid cognition, which remains untouched by this procedure, is the knowledge of the *cogito*, which thereby becomes the absolutely evident beginning of philosophy. (4) The *cogito* bears intentionally the world as a *cogitatum* in itself, and on this basis Husserl insists that the Cartesian path to the reduction grasps pure subjectivity in its full range.[3]

---

[2] *Ideas I* is not the only text, which spells out the *Cartesian* path to the reduction. *Die Idee der Phänomenologie*, the first part of *First Philosophy II* and *Cartesian Meditations* are the other three main texts in which this path is thematized. Although I will concentrate on how this path is thematized in *Ideas I*, my critique just as much applies to how the Cartesian way is presented in these other works.

[3] See Kern (1962).

Thus to characterize this path as Cartesian, as Husserl himself does, is to suggest that it is guided by the principle of *absolute justification*, which assumes the form of *apodictic indubitability*. Being guided by such principles, Husserl in *Ideas I* finds himself compelled to distinguish between the absolute being of the immanent and the merely phenomenal being of the transcendent (Hua III/1, §44), and to assert absolute consciousness as the *residuum* which remains unaffected by the notorious experiment of the annihilation of the world (Hua III/1, §49). Absolute consciousness is thereby identified as the phenomenological field, which is established as apodictically evident. If one is to disclose the "transcendental life" of this consciousness and show how it entails a horizonal structure, absolute consciousness needs to be further broadened so as to entail its past and its future. *Yet such a broadening*, Husserl's claims in *Ideas I* to the contrary notwithstanding, *cannot be from the start established as apodictically indubitable*.[4]

In *Ideas I* Husserl, of course, claims to have established subjectivity as the stream of consciousness in its full temporality. Yet it is hard to take these affirmations at their face value and even harder to find a phenomenological justification for them. In contrast to these affirmations, in *First Philosophy II* and the texts which have been published as appendixes to these lecture-courses, Husserl openly declares that the Cartesian path does not clarify how pure consciousness manifests its own temporality; this path only naively presupposes that consciousness must be temporal (Hua VIII, 433). This is even more clearly expressed in Appendix XIII (1929) to Walter Biemel's edition of *Ideas I*, where Husserl poses a rhetorical question: "have we not constantly presupposed that we have a stream of lived-experience—an endlessly streaming movement of pure life; and that I, the perceiver, do not just perceive this and every other lived-experience (*Erlebnis*), but that I also have a unitary experience (*Erfahrung*) of my life, which is given to me indubitably" (Hua III, 399)?[5]

Following Husserl's own lead, I would like to suggest that the Cartesian path does not establish subjectivity, taken in its full temporal manifestation, as apodictically certain; this path only presupposes that subjectivity is temporal. Yet one should not overlook that when Husserl recognizes this limitation of the Cartesian path in *First Philosophy II*, he does not take it to mean that the idea of apodictic evidence must lose its preeminent significance. According to Husserl, to recognize the Cartesian path as insufficient is to postpone the apodictic establishment of full subjectivity

---

[4]Precisely here Husserl locates the limits of the Cartesian path to the reduction in *First Philosophy II*: "Die mögliche epochē hinsichtlich der ganzen Erfahrungswelt macht in einer auf das erkennende Subjekt gerichteten Reflexion die reine Subjektivität sichtlich, die verbleibt, auch wenn die Welt nicht wäre. Diese Subjektivität wurde dann über die Gegenwart in Vergangenheit und Zukunft hinein erweitert, zum reinen Subjekt im beiderseitig endlosen transzendentalen Leben. Aber was sich so ergab als reine Subjektivität war noch nicht als apodiktisch zweifellos herausgestellt." (Hua VIII, 126)

[5]"Haben wir nicht immerzu vorausgesetzt, daß wir einen Erlebnisstrom—einen endlos strömenden Zug eines reinen Lebens haben, daß ich, also der Wahrnehmende, nicht nur dieses und jenes Erlebnis wahrnehme, sondern eine *einheitliche Erfahrung von meinem Leben habe* von der her ich desselben zweifellos gewiß bin?" (Hua III, 399)

"until later." Husserl thus claims that the reduction must first be disclosed as *phenomenological* and *transcendental*, and only afterwards submitted to an *apodictic* critique (Hua VIII, 80, 126). It is curious, yet no less telling, that as this critique is put aside, Husserl never returns to carry it out.[6]

Such a state of affairs presents us with an awkward alternative. On the one hand, if we choose to establish the domain of phenomenology as a domain that is from the very start marked by indubitable givenness, we need to give up the prospects of obtaining full subjectivity. On the other hand, if we choose to thematize full subjectivity, we need to give up the principle of apodictic indubitability, at least as far as it refers to the necessity of beginning with absolute certainty. *From within the strict boundaries of the phenomenological field opened up by the Cartesian path to the reduction*, one is to assert that horizon-consciousness, as thematized in Sect. 6.4, and subsequently the full subjectivity, which I addressed in Sect. 6.5, cannot be considered a phenomenological given since both lie outside the domain, as the latter is established within static phenomenology. *Apodictic indubitability as a methodological Leitprinzip, which demands that phenomenology's beginning consist in absolute evidence, suppresses the phenomenality of the horizon.*

We thereby obtain an answer to the question why the genetic phenomenality of the horizon remained unthematized in static phenomenology in general, and in *Ideas I* in particular. We also see that if the horizon-consciousness is to be considered a phenomenological theme at all, it must be established on the basis of a different path to the reduction.

7.3 The Double Reduction: The Discovery of the Horizons of Consciousness and the Emergence of the Full Subjectivity

Husserl's thematization of the new path to the reduction in *First Philosophy II* validates the proposed thesis that the disclosure of transcendental subjectivity goes hand-in-hand with the disclosure of the horizons of consciousness. Of central significance is the notion of the *double reduction*, first mentioned in this work in §39. This notion signifies a shift in perspective from within which the reduction is thematized in phenomenology. Instead of following the Cartesian path, which in one blow suspends the world-thesis and thereby brings one into the phenomenological field, in *First Philosophy II* Husserl thematizes the reduction not as it pertains to the world-thesis, but rather as it pertains to the appearance of concrete objectivities.[7]

---

[6] I will return to this issue in Sect. 7.8.

[7] "Hier aber liegt folgender Gedanke für einen neuen und scheinbar einfacheren Weg zur transzendentalen Subjektivität nahe: Genügt es nicht, ohne mit jener langwierigen Kritik der Welterfahrung zu beginnen und die Möglichkeit der Nichtexistenz der Welt zur Evidenz zu bringen, direkt an den einzelnen Akten die epochē des interesselosen Selbstbetrachters ins Spiel zu setzen?" (Hua VIII, 127)

Particular objectivities are given not only through perception, but also through remembering and expectation. Yet the manner in which these objectivities are transformed into transcendental phenomena is different in the case of perception, on the one hand, and in the cases of remembering and expectation, on the other. While in the case of perceptual acts, this transformation is one-dimensional in that it calls for the suspension of all fulfilling apperceptions along with the suspension of the already fulfilled perceptual belief, in the remaining two cases, the phenomenological transformation calls for a *double reduction*.

This double reduction in its first phase is identical with the reduction as it pertains to perceptual acts. To reduce a remembrance of a particular objectivity is to inhibit the experiential belief and thereby to obtain remembrance as a *present* phenomenological experience. This present experience, as recollection and expectation show, already entails appearances not only as present, but also as past and future. We thereby see that the experience of the now is in possession of horizons; yet the latter are exclusively *horizons of objectivities*, and not of the *intending* of objectivities.

And yet, "I remember" also entails "I have perceived," or in the case of action, "I have done so and so." Here we encounter a limitation that pertains to the first phase of the reduction. The first reduction leaves the *content* of my past experience and actions out of account; and it does so on the grounds that what is taken to be an actual recollection could in truth have never taken place.

The possibility that each particular experience could be nothing more than a transcendental illusion calls for the suspension of *objective* time. Yet such a suspension remarkably leaves the *continuity of my experience* intact: even if my remembrance does not guarantee that an experience in question truly took place in the objective past, nonetheless, since there is no remembering without the remembered, the experience in question is already stamped as having been a past experience. Of central importance is the realization that with the reduction, the *content* of remembrance need not be placed out of action. The reduction, besides releasing the "I remember" as a present transcendental experience, can also entail its content as a "piece" of my *transcendental* life. Emphasizing that this structure pertains to all remembering, Husserl insists that "every remembrance *evidently* [*evidenterweise*] authorizes a double transcendental reduction, the first of which yields my present transcendental experience, while the second one, curiously reaching into the reproductive content of remembrance, reveals a piece of my past transcendental life" (Hua VIII, 85 [*my emphasis—SG*]).[8]

What sense are we to make of Husserl's employment of the notion of *evidence* in this sentence? Does the presence of this word not indicate that the double reduction, which *First Philosophy II* announces, could just as well have been established in the

[8] "Jede [Erinnerung] läßt evidenterweise *eine doppelte transzendentale Reduktion* zu, deren eine die Erinnerung als mein transzendentales gegenwärtiges Erlebnis ergibt, während die zweite, in merkwürdiger Weise in den reproduktiven Gehalt der Erinnerung hineingreifend, ein Stück meines vergangenen transzendentalen Lebens enthüllt" (Hua VIII, 85). See also Hua VIII, 129, 130, 159.

confines of the Cartesian path to the reduction? That is, could the givenness of a "piece" of my past transcendental life not be said to be an *apodictic* givenness? Husserl directly claims that such indeed is the case in Appendix VIII to Hua XI, significantly entitled "The Apodicticity of Remembrance" ("Die Apodiktizität der Wiedererinnerung" [Hua XI, 365–383]). And yet, to procure a "piece" of my past is not yet to reveal consciousness as a stream; nor is it, for that matter, to account for the full significance of the double reduction. A closer look reveals that the appearance of "pieces" is meaningful only as far as they are "pieces within a chain," and if I choose to follow these chains, each of my past experiences can lead me back to my present and thereby disclose my transcendental past in its continuity.[9] Moreover, each past experience can motivate me to inquire into what precedes it, and if I follow the "chains" of these awakenings, I disclose my transcendental life as continuously leading back to its, as Husserl emphasizes, *endless* past.[10]

Let us turn back to the question of the "chains" of experience. The disclosure of what such "chains" entail *cannot be given apodictically*, as Husserl has already realized in his lectures, *Die Grundprobleme der Phänomenologie* (1910–11). Not only does the manner in which a past experience relates back to the present entail a myriad of suppositions and hypotheses; it also entails a sundry of temporal spans, which can only be given emptily. At best, the continuity of the transcendental life is *implied* in the double reduction; and thus its apodictic evidence, as Husserl directly claims in Appendix XIII to *Ideas I*, is an *Idea* whose full determination lies in infinity (Hua III, 401). Thus even if one were to agree that the double reduction discloses a "piece" of my transcendental past apodictically, it nonetheless cannot lead to the establishment of the continuity of my transcendental life as apodictically certain. Husserl's analysis of the double reduction thereby corroborates the claim that *one needs to choose between apodicticity as a phenomenological ground and the phenomenality of the horizon*.

One could, however, object that even if the presumably apodictic givenness of a "piece" of my past does not establish the continuity of my transcendental past as apodictic, this does not yet mean that the apodicticity of the latter could not find its grounds elsewhere. Thus in the context of discussing the path to the reduction spelled out in *First Philosophy II*, Antonio Aguirre suggests that,

> It becomes clear that no act of the subject is an isolated unity, but rather each act, *from the very start*, is a dependent moment interwoven into the web of the continuity of life.

---

[9]Consider in this regard Husserl's analysis of the past horizons in the C-Manuscripts. After indicating that the past that memory reawakens is itself given with a futural horizon, Husserl continues: "Zu diesem [Horizont der vergangenen Gegenwart] gehört eine Zukunftsstrecke, der meine Gegenwart und ein Zwischen ihr und jener Vergangenheit angehört, ein Zwischen, das schon vergangen ist und für die vergangene Gegenwart noch (unerfüllte) Zukunft ist, jetzt aber für mich in Form jenes 'Zwischen' erfüllte, erledigte Künftigkeit jenes Vergangenen." (Mat. VIII, 44)

[10]The endlessness in question is to be understood as a *negative* claim, i.e., as a claim that my past is never given to me in such a manner as though it had a clearly defined starting point. Far from it: Each recollection is given to me as a fulfillment of an expectation. Within such a context, the question of the first experience proves senseless, and in this sense, my past reveals itself to me as limitless, i.e., as endless.

> An epochē that is applied to a singular act cannot suffice, since each singular turn already presupposes the universal turn. And so psychology, which *still* reflects in the natural context, can only begin with the universal reduction; the universality appears at the beginning and not at the end of reflection (*Besinnung*), in respect to the application of the epochē. (Aguirre 1970, 47)[11]

Admittedly, the transition from the psychological to the philosophical reduction is highly problematic, as the appendixes to *First Philosophy II* make clear. Yet despite these difficulties, Husserl's analysis brings to light that the universality of transcendental life does not precede, but rather follows what the double reduction establishes. If this more rudimentary universal reduction is to be seen as the one disclosed through the Cartesian path, then in contrast to Aguirre I find myself compelled to insist that this path, as the previous section has shown, fails to disclose subjectivity as temporal. It is by far not clear how much support an attempt to disclose such deeper lying sources of universality can find in Husserl. As Husserl remarks in 1929,

> Admittedly, the structure of *apodicticity* of my own being, as the being of the pure I that belongs to my pure life, and this life itself in its relation to the temporal, immanent-temporal whole of my being—none of this is without difficulties. For instance, the immanent remembrance can very well deceive me, and thus conflict, deception, being-otherwise (than the intuitive remembrance shows) is clearly possible beyond the boundaries of the living intuitive immanent presence. (Hua III, 401)[12]

Thus the principle of apodicticity cannot be seen as a presupposition underlying the universality of transcendental subjectivity; apodicticity at best can be conceived as an *Idea*, which remains to be established after disclosing the horizons of subjectivity.

Just as our experience of the past, so our experience of the future also calls for a double reduction. And even if the future does not present itself as strictly analogous to the experience of the past—after all, what I anticipate is not given as determinately as what I remember—nonetheless, the general structure of the reduction that pertains to anticipation and to remembrance is largely the same. Here also we face a "double transcendental"—on the one hand, expectation as a present experience, and on the other hand, the entailed content of expectation. And just as the experience of the past opened up its chains through which it revealed itself as endlessly continuous, so expectation also entails such chains, which lead me to my *endless* future.

---

[11] "Kein Akt des Subjektes, so wurde klar, ist eine isolierte Einheit, sondern jeder ist *von vornherein* als unselbständiges Moment ins Netz des Lebenszusammenhanges verflochten. Eine auf einzelne Akte angewandte Epochē kann es demnach gar nicht geben, sondern die Einzelanwendung setzt die universale schon voraus. Auch der Psychologe, der *noch* innerhalb der Natürlichkeit reflektierende, kann also nicht anders als mit einer universalen Reduktion anfangen; die Universalität steht am Anfang und nicht am Ende der Besinnung bzw. der Anwendung der Epochē." (Aguirre 1970, 47)

[12] "Nun macht freilich die Struktur der *Apodiktizität* meines Seins als des reinen Ich meines reinen Lebens und dieses selbst auf das zeitliche, immanent-zeitliche Ganze dieses Seins und Lebens ihre Schwierigkeiten. Da doch z.B. immanente Erinnerung sehr wohl täuschen kann, also Widerstreit, Täuschung, Anders-sein (als selbst anschauliche Erinnerung zeigt) sehr wohl möglich ist außerhalb der lebendig anschaulichen immanenten Gegenwart." (Hua III, 401)

Yet what exactly does it mean to suggest that the content of each and every remembrance, as well as of anticipation, is a "piece of a chain" which entails within itself references to other lived-experiences? The metaphor of a chain—and let us for the moment suspend the question of its appropriateness—indicates that *each experience of the past and of the future is given within a horizonal framework which itself is nothing other than my transcendental life*. As Husserl has it, "provided that *every* presence always carries with it a future horizon of expectation, we have again, analogously to the endless horizon of the transcendental past, an *open endless horizon of transcendental future*" (Hua VIII, 86).[13] These horizons reveal themselves as *horizons, which belong exclusively to subjectivity*, primarily because the "chains," in which each and every *experience* manifests itself, make their appearance in the wake of the suspension of objective time.[14] It is the uncovering of these "chains" that brings forth, as the title of §39 of *First Philosophy II* suggests, the full content of universal self-experience, which is nothing other than the content of the full subjectivity of which I spoke in Sect. 6.5.

Let us call the following characterization of the horizon its first determination that *First Philosophy II* brings forth: *the horizons of subjectivity reveal themselves as the "mirroring," or the implication of the whole life of consciousness within each lived-experience*. It is hard not to see this realization as Husserl's secular alternative to William James' remark in his *Principles*: if the divine mind could glance into any of our experiences, he would see within each of them the whole life of subjectivity. Although the horizons subjectivity inhabits are manifold, each of them ultimately merges with all others, thereby necessitating one to speak of the *horizon* of subjectivity in the singular. At this stage we are not yet ready to ask how the horizonal framework of subjectivity limits the range of possibilities that pertain to experience, even though this question already lurks in the background. We are still in need of gaining a better understanding of the structure of horizonal references. We find Husserl's further-reaching thematization of how each experience entails references to other experiences under the heading of "intentional implications and iterations."

7.4 Intentional Implications as a Method for Thematizing the Transcendental Life of Subjectivity

The horizons under scrutiny are horizons of subjectivity not only because they are established after objective time is placed in brackets, but also because each reference to other experiences that the horizons entail *implies a content that belongs to*

[13]"Und sofern immerzu *jede* Gegenwart einen Zukunftshorizont der Erwartung kontinuierlich mit sich führt, haben wir wieder, analog dem endlosen Horizont der transzendentalen Vergangenheit, einen *offen endlosen Horizont einer transzendentalen Zukunft*." (Hua VIII, 86)

[14]"Wir sehen, mit der Welt ist die *objektive Zeit*, diejenige Zeit, welche die Form der mundanen Objektivitäten als seiender ist, außer Spiel gesetzt. Andererseits aber: Ich, das transzendentale Ich, lebe ein transzendentales Leben [...] das in sich den endlosen Horizont der Erinnerung und Erwartung trägt, der, enthüllt, einen beiderseits endlosen transzendentalen Lebensstrom zeigt." (Hua VIII, 86)

subjectivity itself. That is, the content of the past and the future acts is the content of experiences that either belonged to or could belong to the past or the future ego. Thus the implication of the past and the future experiences is also an implication of the past and future ego. Now the notion of the *implicated* ego is revealing in that it shows that the ego in question is not the same as the one having the particular acts of remembrance or expectation. That such is the case becomes evident when the notion of intentional implications is employed not only in reference to the acts that provide us with the knowledge of the past and the future, but also in reference to phantasy. Here also, each and every object of imagination intentionally implies an ego within the phantasy world, which is clearly distinct from the ego having the particular phantasies (Hua VIII, §44). Analogically, the same must be said about the ego implicated in recollection and expectation.[15] Yet there is a crucial difference between phantasy on the one hand, and remembrance and expectation on the other, since "every appropriate reflection tells me: I, being the same I that I remember myself to be, was already there at the time which I now remember" (Hua VIII, 134).[16] And as we have already seen, the ground for this certainty stems from the fact that each past or future *Erlebnis* is a "piece within a chain" which I can follow through its intentional implications until it merges with the present ego.

The notion of intentional implications enriches the manner in which subjectivity reveals itself in its horizonal extension as addressed in Sect. 7.3 in that it enriches the notion of the double reduction in such a way that *it is shown not to be robust enough to reveal subjectivity in its full manners of manifestation*. Of crucial significance is the realization that what is implicated in the acts of recollection and expectation is open to further iteration. The content of my remembrance can itself be another remembrance.[17] And just as in place of a straightforward remembrance, I can have a memory of another memory, so in place of a simple expectation, I can have an expectation of an expectation. These chains of intentional implication can stretch into infinity; and each of the "pieces within the chain" can be given within the phenomenological field if, and only if, each of them undergoes a reduction. Thus instead of a double reduction, the notion of intentional implications invites one to speak of reductions that stretch into infinity.

Yet even this notion of the reduction does not yet take into account all that is entailed in the notion of intentional implications. It should not be overlooked that different kinds of acts can be implicated within memory or expectation. Thus I can have a memory of an expectation as well as an expectation of a memory, a phantasy of an expectation as well as an expectation of a phantasy, etc. What I said in regard to the chains of recollection and expectation remains true in regard to these complicated

[15] "Erinnere ich mich, so bin ich, das erinnernde Ich, als Subjekt des Aktes natürlich nicht im intentionalen Gehalte des Aktes. Aber gleichwohl—in einer Modifikation, in der meiner subjektiven Erinnerungsvergangenheit, bin ich doch zu ihm gehörig." (Hua VIII, 134)

[16] "Jede passende Reflexion sagt mir: Ich, derselbe, der ich jetzt mich erinnere, war beim Erinnerten dabei." (Hua VIII, 134)

[17] See Hua VIII, 136.

chains of acts as well: each of them can be considered a phenomenological given only when each of them is placed under the reduction.

The notion of intentional implications enriches the already provided account of the reduction not only in regard to the *content* given in each and every particular experience. *The intending of this content can also be reiterated.* Just as in the case of any act I can bracket the object and retain it only as it is perceived, remembered, expected, or phantasized, so I can also bracket the object of the *reflecting phenomenological act* and thereby retain only the act itself. Nothing stands in the way of my progressive enrichment of the chains of the *epochē* and thus of the thematization of the reduction that would belong to the second, then third, then fourth, etc. level.

For Husserl, such a possibility indicates the *impossibility of the last reduction*: "and so we reach in any iteration at least the ideal endlessness of phenomenological reductions" (Hua VIII, 138).[18] One should not, however, be too quick at identifying this proclamation with Merleau-Ponty's famous announcement that "the most important lesson which the reduction teaches us is the impossibility of a complete reduction" (Merleau-Ponty 1976, xiv). For Husserl, the admission of the impossibility of the last reduction does not bespeak the finitude that belongs to each and every human endeavor or the need for a progressive deepening of the reduction so as to counteract the limitations, which inevitably pertain to its present form. For Husserl, such an interpretation would amount to a misunderstanding of phenomenology's central intentions, and so as to avoid it, he remarks: "what I have exposed as the purely subjective in the phenomenological experience neither loses nor wins anything when I, by means of a repeated reflection, become aware that the phenomenological exposition is itself an act and that I, being interested in it, so as to expose it in pure experience as a pure act, must reduce this act at a higher level" (Hua VIII, 138).[19]

The notion of intentional implications enriches the characterization of the horizon of subjectivity as the implication of the whole life of consciousness within each lived-experience. This characterization does not only mean that each *Erlebnis* has a temporal position through which it implicates all other *Erlebnisse* in terms of a twofold temporal extension. Enriched by the notion of intentional implications, this characterization also takes into account the unique features that pertain to *Erlebnisse* themselves, features through which the horizonal reference no longer appears as merely two-directional, but as extending in manifold directions of remembrance, expectation, and phantasy, when the latter are taken along with all possible modalities that relations between them generate. Thus the whole life of transcendental subjectivity can no longer be represented in terms of a linear expansion; it is rather the case that each *Erlebnis* shoots its projections in manifold directions, each of which is entailed

[18] "So kommen wir in beliebiger Iteration zu mindestens ideellen Unendlichkeiten von aufgestuften phänomenologischen Reduktionen." (Hua VIII, 138)

[19] "Was ich nämlich als rein Subjektives in phänomenologischer Erfahrung herausgestellt habe, verliert und gewinnt nichts dadurch, daß ich nun wieder reflektierend dessen inne werde, daß das phänomenologische Herausstellen selbst wieder ein Aktus ist und daß ich, dafür interessiert, ihn als reinen Aktus in reiner Erfahrung herauszustellen, eben wieder auf ihn, diesen Akt höherer Stufe, eine Reduktion beziehen muß." (Hua VIII, 138)

within the horizon of subjectivity. Here we see why the metaphor of a "chain" is not robust enough to capture the complexity that pertains to the manifold ways through which each *Erlebnis* relates to all other *Erlebnisse*. What we face here is rather an intertwining of numerous "chains," all of which form an integrated whole.

7.5 The Paradoxical Relation Between the Reduction and Horizonality. The Emergence of the Universal Reduction

Yet with such a notion of horizon-consciousness, we stumble into an awkward situation. On the one hand, it was the abandonment of the requirement to begin with the reduction, as a method, which from the very start brackets the world-thesis, that enabled phenomenology to thematize horizon-consciousness. Yet on the other hand, the very fact that the world-thesis has not been placed into brackets leads to the inescapable recognition that the horizon-consciousness, as thematized so far, is not yet, in the strict sense, phenomenological.[20] The reason for such a curious situation lies in the fact that *the horizon is not only a horizon of reference*, but also a *Geltungshorizont*, a horizon of validity. This is *the second sense* in which *First Philosophy II* thematizes the horizonality of experience. And even if this notion is not as thoroughly analyzed in *First Philosophy II* as the first notion, nonetheless, it is no less significant.

If validities of sense, which *Geltungshorizont* entails, are already operative before the suspension of the world-thesis, then they do not merely indicate the achievements of consciousness (*Bewußtseinsleistungen*), but also are components, which lead back to the unquestioned, and in this sense philosophically unjustifiable, worldly apperceptions.[21] These apperceptions indicate that the foregoing analysis of the reduction is in truth not *philosophical*, but rather *psychological* in that it presupposes the objective apperception "a human being" as its basis and therefore builds its accomplishments as *components* of this objective apperception.[22]

[20] Here the notion "phenomenological" refers to the specifically *philosophical* attitude. Borrowing a phrase from Antonio Aguirre, one can characterize horizonality, as it has been so far unfolded, as a notion, which belongs to the "phänomenologische Natürlichkeit." (Aguirre 1970, 46)

[21] "Daraus folgt, daß, wo immer ich in objektiver Verflechtung mit einem schon reduzierten Objektiven noch einen Rest übrig habe, an dem nicht Reduktion geübt worden ist, also einen Bestand an Objektivem, das schlechthin gesetzt ist, statt im Rückgang auf das Bewußthaben von ihm als bloß intentionale Gegenständlichkeit dieses Aktes gesetzt zu werden, ich noch nicht ein wirklich und vollkommen reines Subjektives habe." (Hua VIII, 139) "So führt jedes rein Subjektive, das ich reduzierend gewinne, beständig und von dieser Reduktion unbetroffen eine *objektive Geltungskomponente* mit sich, die aus jener *Verflechtung* der objektiven Geltungen, der nie inhibierten, herstammt." (Hua VIII, 141)

[22] And thus Husserl insists that "das phänomenologische Verfahren, das wir bisher an verschiedenen Typen von Einzelakten geübt hatten, leistet nichts anderes und kann nichts anderes leisten als *die seelische Innerlichkeit in ihrer Reinheit* herauszustellen und die in ihr verborgenen *intentionalen Implikationen* zutagezufördern." (Hua VIII, 141)

Thus on the one hand, the absence of the reduction of the world enabled us to uncover the horizons of consciousness; yet on the other hand, this absence subsequently reveals that the horizons are not yet fixed as horizons of subjectivity. Or better, it is these very horizonal implications, established in the absence of the world's annihilation, that reveal themselves as still entailing mundane components. The recognition that the horizons are horizons of validity and, furthermore, that the horizons of validity operate as *habitualities of validity* (*Geltungshabitualitäten*), is nothing less than an admission that "the real world has (for me) validity, taken along with all of its experiential actualities and some ideal actualities, which I have, in my past life, come to know as existent (als seiend) and which in my realm ... remain intact" (Hua VIII, 143).[23] Thus if releasing the horizons of subjectivity can be considered the fundamental accomplishment of the new path to the reduction that *First Philosophy II* spells out, then this releasement itself leads to the recognition of a fundamental shortcoming of the path that disclosed them. The psychological path to the reduction unfolds the horizonal structure of consciousness, yet in such a way that these horizons subsequently reveal the inadequacy of this path.

Yet the paradox we face here does not lead to a dead-end, for nothing stands in the way of the performance of a *universal* epochē *after* the disclosure of the horizonal framework of experience. However, in contrast to *Ideas I*, in *First Philosophy II* the universal *epochē* and the reduction are not thematized in terms of the suspension of the world-thesis but rather in terms of *inhibiting of all worldly interests* (§41–§43).[24] Although so far I have kept the notion of *interest* out of considerations, it nevertheless lurked in the background in the discussion of the double reduction and of intentional implications. For what else can it mean to speak of reduced phenomena if not of their givenness, which is secured by shifting our attention, i.e., our interest, from their natural to their subjective manifestation? Yet the paradox that engulfs the relation between reduction and horizonality makes clear that the shift of interest in regard to concrete objectivities is by far not sufficient. Such *disinterestedness* suspends only *actual* validities and fails to inquire into *habitual* validities that, be it more anonymously, yet nonetheless just as effectively, stamp appearances as worldly.

Thus we are in need of a universal *epochē*, conceived in terms of a *universal* shift of interest. The notion "universal" here indicates that the *disinterestedness* in question inhibits not only actual but also habitual validities. To place the latter out

[23] "Ich habe also meine Stellungnahme zum wirklichen Sein der intentionalen Gegenstände der jeweiligen Akte nicht schlechthin und absolut inhibiert, und erst recht nicht meine sonstigen Seinssetzungen, meine gesamte Geltungshabitualität, durch die für mich beständig die reale Welt da ist, Geltung hat, mit allen ihren erfahrungsmäßigen Wirklichkeiten und auch mancherlei idealen Wirklichkeiten, die ich in meinem früheren Leben als seiend erkannt habe und die in meinem Reich von Erkenntnisbesitztümern (habituellen Überzeugungen) unberührt erhalten bleiben." (Hua VIII, 143)

[24] In this context, Husserl speaks of disinterestedness. Yet it must be kept in mind that this notion does not mean an abandonment of each and every possible interest, but rather the releasing of the *subjective* interest: "*Es ist reines Interesse am subjektiven Sein.*" (Hua VIII, 108)

of action does not mean to "reduce" validities to mere phenomena; rather, it means to retain them as validities of sense, which do not manifest themselves as appearances, but as *validities* that encircle appearances. Moreover, it ultimately means to recognize that they have their origin in the positing, enacted by consciousness itself.

My brief consideration of the universal reduction, as employed in *First Philosophy II*, shows that this notion does not suppress the horizonal framework of consciousness the way the universal reduction in *Ideas I* did. On the contrary, *only through* such a broadening of the reduction do the horizons reveal themselves as *horizons of subjectivity*. Yet also, through such a broadening, the very sense of horizons as subjective undergoes a significant transformation. When in this chapter I spoke of the horizons as the implication of all the life of consciousness within each *Erlebnis*, I meant by this only an intentional reference that each and every *Erlebnis* entails to all other *Erlebnisse*. And even though I made it clear that such a characterization is not sufficient once interpreted in terms of a two-directional temporal extension, nonetheless, even in its more precise form, such a conception of horizonality remains insufficient. The universal reduction requires that we supplement this characterization of horizonality by way of linking it to the characterization of the horizon as *Geltungshorizont*. The universal reduction requires one to say: *the horizons are horizons of implicit validity which embrace appearances as "transcendence within immanence;" this inseparable dimension of sense which belongs to appearances, even though it is not of appearances, is nothing other than the implication of all the life of consciousness within each and every lived-experience.*

7.6 Horizon-Consciousness as a System of Implications and the Source of Validity

The merging of the two notions of horizonality has so far been presented only on the methodological level. What I have shown so far is only the methodological necessity to synthesize these two notions so as to secure the horizon as the horizon of subjectivity. From §49 of *First Philosophy II*, which directly follows Husserl's analysis of the universal reduction, one can derive what such a merging means on a *thematic* level. That is, following the analysis provided in this section, one can show what the merging of these notions means for all *appearances* that enter the field of consciousness. Let me remark that even though in *First Philosophy II* the notion of the horizon figures relatively often in the context of thematizing the second path to the reduction, not until §49 does it become the focus of Husserl's investigations. Thus the merging of these two notions of horizonality is operative within Husserl's direct analysis of the horizon *from the very start*, albeit no conceptual distinction between these notions is explicitly drawn.

Here the inner- and the outer-horizons, which I have already addressed in the first part of this work, are at the center of Husserl's attention. It is significant that the analysis of these fundamental types of horizonality is from the very start interpreted as horizonality that belongs to consciousness itself. At first glance such a claim appears counter-intuitive: do the horizons in question not belong to objectivities? Yet a closer look reveals that the emptiness, which the notion of horizonality bespeaks,[25] pertains to the *non-intuitive awareness of the non-actual manners of objectivity's givenness*; or as Husserl has it, "this intuitively empty consciousness is still consciousness, a horizon of consciousness" (Hua VIII, 146).[26]

Husserl begins his analysis of the inner- and outer-horizons by disclosing the implicit references entailed in each mode of appearance. Thus each appearance is said to have its inner background, whose possible manners of givenness are entailed in the present appearance. And even though the side unseen lies outside my perceptual field, and thus a simple shift of attention does not provide me with its givenness, nonetheless, I can always, by modifying my orientation within the perceptual field, render it visible. Now while the inner-horizon is by definition non-intuitive, the outer-horizon entails both an intuitive and a non-intuitive dimension of co-givenness.[27] That is, a thematic object, as it is given in the perceptual field, is given as an object in the midst of other objects. Yet the perceptual field does not exhaust all co-givenness: the book that is given to me is on the desk, is in one of the library rooms, which itself is a building in the city, etc. And even though this spatial *plus ultra* is not given explicitly in expressive acts of consciousness, nonetheless, the intentional references to what lies outside the perceptual field can always be drawn.

So far my account of the inner- and outer-horizons can be understood as an exemplification of the horizon, conceived only in terms of intentional references. Yet Husserl quickens to add: the horizon denotes "a unity of validity, which entails various particular validities that can be exposed" ("eine Einheit der Geltung, mit vielerlei herauszustellenden Sondergeltungen" [Hua VIII, 147]). Indeed, the reference to objectivity's other modes of appearances through its inner-horizons co-determines the *sense* of the *thematic* appearance in that *only through such references* can an appearance be conceived as an appearance of a particular objectivity. Such co-determination precisely indicates that the horizon functions as a source of *validity*. Even my assurance that what I see in front of me as a particular object already entails a projection of *validity* which pertains to the manner in which this particular object, of which I see only a part, is given through its other modes of appearance. The same holds true of the outer-horizon: the sense of what confronts me is co-determined by the *validity* that stems from horizonal co-givenness.

Thus inner- and outer-horizons are possible only so far as the horizons in question entail not only the dimension of implicit reference but also of implicit validity.

---

[25]The emptiness in question is not the emptiness of anticipation, but the emptiness of intuition: "Völlig sinnesleer ist aber selbst das unbestimmteste Bewußtsein hierbei nicht, es ist zumindest ein Raumding, ein Farbiges etc. gemeint." (Hua VIII, 146)

[26]"Dieses anschauungsleere Bewußtsein ist doch Bewußtsein, ein Bewußtseinshorizont." (Hua VIII, 146)

[27]"Vielmehr scheiden wir im gesamten Außenhorizont den Bereich noch wahrnehmungsmäßiger Anschaulichkeit und den unanschaulichen Leerhorizont, der sich kontinuierlich anschließt." (Hua VIII, 147)

Only so far as the two notions of horizonality are taken together can the horizon entail the dimension of a preliminary indication (*Vordeutung*) and of the possibilities of experience; and only so can the horizon-consciousness be seen not merely as a notion that embraces empty possibilities of imagination but rather as a notion which bespeaks what lies at the heart of positional consciousness itself. The horizon as the *unity of validity* (Hua VIII, 147) indicates the *homogeneous style of the forward streaming experience*. The form of validity, which the horizon announces, is the form of manifold possibilities which, despite their apparent endlessness, are in no way merely possibilities of phantasy.

7.7 Towards a New Way of Questioning the Origins of the Horizon

We have obtained the unity of the two notions of the horizon on both the methodological (Sect. 7.5) and the thematic levels (Sect. 7.6). Yet if one where to ask whether *First Philosophy II* discloses the far-reaching consequences that pertain to the unity of these two notions, one could not ignore Husserl's own answer: "if we wanted to intrude at a deeper level, a great field of ever new studies would open up" (Hua VIII, 150).[28] Husserl further indicates three possible directions, in which a further-reaching investigation of horizonality could be pursued.

First, one could inquire into the manner in which the persisting validity originates in the originary founding acts; how, moreover, a particular validity comes into harmony or conflict with other validities; how, finally, the world comes to be seen as the title for the total stock (*Gesamtbestand*) of positive validities. Secondly, one could also inquire into the structure of the pregiven world itself, which, despite the changes that pertain to particular convictions, always remains in its universal structure what it already is—the world as a spatio-temporal and causal nature, which is already inhabited by animals and humans, who relate to each other through social ties, who already are in communities or societies. Such universal structures of the existing world, of which Husserl speaks in terms of the *world-horizon*, are always already pregiven before any kind of practical engagement in the world takes place.

Yet the world-horizon entails not only real, but also *ideal* objectivities, and therefore, thirdly, one can also inquire into the manner in which the ideal "worlds," composed of ideal objectivities, come into being. These ideal worlds also, once formed, belong to our habitual resources, which, just as everything else habitual, can be inquired into in terms of the reactivation of their origins.

These possibilities are only mentioned and in *First Philosophy II* they are not pursued any further. Husserl engages directly in these questions in *Analyses Concerning Passive Synthesis* and *Experience and Judgment*. Yet before turning

[28] "Wollten wir tiefer eindringen, so eröffnete sich hier ein großes Feld immer neuer Studien." (Hua VIII, 150)

to these analyses in the next chapter, I want to emphasize that the central significance of these questions relies upon the fact that they reveal *the horizon as a central concept within genetic phenomenology*. That is, these possibilities reveal the horizon as the milieu of concealed sense accomplishments, *as the very field of sense which genetic phenomenology aims to thematize*. The indicated possibilities thereby tie the horizon of subjectivity to the phenomenological inquiry into origins with an inseparable bond. More precisely, they reveal that the phenomenological thematization of the origins of sense is an inquiry into the formation of horizonality. *The horizon of subjectivity thereby proves to be the thematic field of genetic phenomenology itself, for to thematize the origins of sense-formation is nothing other than to delineate the crystallization of the horizon.*

We thereby gain another perspective from within which we can address anew our main theme—the origins of the horizon in Husserl's phenomenology. When I addressed this question in the first part of this work, my central concern pertained to the emergence of the horizon as a philosophical notion and a philosophical theme. Now we already see that to inquire into the origin of the horizon can also mean to unfold the conceptual framework within which an inquiry into origins itself becomes meaningful. *To describe the horizon as the hidden space of rudimentary sense-accomplishments is nothing other than to uncover its genetic phenomenality.* We are ready to ask: how does the thematization of rudimentary sense-formation enrich our understanding of the horizon? I will turn to this question in Chap. 8.

7.8 Landgrebe's Interpretation of First Philosophy II: The Question of Cartesianism

My foregoing analysis suggests that *First Philosophy II* marks a significant transformation in Husserl's phenomenology. Such a view is by no means unprecedented. It has been in wide circulation since Ludwig Landgrebe's "Husserl's Departure from Cartesianism" appeared in print in 1962, just 3 years after *First Philosophy II* was published in the *Husserliana* edition. To this day, Landgrebe's interpretation remains unparalleled in recognizing the importance of *First Philosophy II* not only for Husserl's phenomenology, but for European philosophy of the twentieth-century in general. I would nonetheless like to suggest that half a century after the publication of Landgrebe's famous piece, the significance of *First Philosophy II* calls for a significant reevaluation.

In "Husserl's Departure from Cartesianism" Landgrebe aimed to provide an indirect response to the growing sentiment that Husserl's phenomenology is a type of Cartesianism, i.e., a type of philosophy, which identifies its central task as that of obtaining apodictic evidence, which in its own turn would serve as the starting point of all genuine knowledge. Such a view of phenomenology was especially prevalent in hermeneutical and existentialist literature, which viewed Husserl's "obsession" with methodological and epistemological issues not only as an innocent reluctance to address the actual problems philosophy was meant to address, but even more significantly, as the most forceful expressions of modern self-estrangement.[29]

As a response to such a widespread sentiment, Landgrebe's task was that of resituating Husserl's phenomenology in the landscape of the current philosophies. Could it not be so that the widespread critical sentiment notwithstanding, Husserl's critics too quickly dismiss his phenomenology? Moreover, could it not be so that the hermeneutical and existentialist contributions are themselves built upon those very insights, which first make their appearance precisely in *First Philosophy II*? Landgrebe aimed to answer both questions affirmatively, and so as to substantiate such a view, he described *First Philosophy II* as a turning point in Husserl's phenomenology. On Landgrebe's view, while Husserl's earlier works aimed to bring modern rationalism to its completion, *First Philosophy II* marks a "departure from those traditions which are determinative for modern thought" (Landgrebe 1970, 260).[30] However, a perspective that introduces a hiatus between phenomenology and modern philosophy as well as sees phenomenology as a precursor to hermeneutics and existentialism could not find support in Husserl's own understanding of phenomenology. Landgrebe realized this all-too-well, and for this reason he further qualified Husserl's departure from Cartesianism as a *reluctant* departure, "insofar as Husserl has wished to complete and fulfill this tradition without knowing to what extent his attempt served to break up this tradition" (Landgrebe 1970, 260).[31] As Landgrebe famously puts it, "metaphysics takes its departure behind Husserl's back" (Landgrebe 1970, 261).[32]

Landgrebe aimed to substantiate such a standpoint on the basis of the role that the concept of *experience* plays in *First Philosophy II*. Landgrebe structures his analysis by asking three questions which, as he himself admits, are nothing else but three aspects of one and the same fundamental concern: (1) In what sense does

---

[29]For a good illustration of such a view, consider F.H. Heinemann's analysis of Husserl in his *Existenzphilosophie lebendig oder tot?:* "Wenn er [Husserl] die Existenz der Welt mit ihren Bergen, Flüssen, Bäumen, Pflanzen und Tieren einklammerte, wie konnte er davon absehen, seine eigene Existenz in diesem spezifischen Körper in Klammern zu setzen? Darum ging er von der menschlichen Einsamkeit zur transzendentalen weiter: 'Ich bin' sagte er, 'zum transzendentalen ego geworden,' und das war tatsächlich der Fall. Er sprach nicht als sein natürliches Selbst, sondern als ein anonymes transzendentales ego, als Bewusstsein überhaupt im kantischen Sinne. Hier haben wir den zentralen Punkt erreicht. Husserls Philosophie ist Philosophie des einsamen, transzendentalen Selbst. Diese Tatsache erlaubt uns, Husserls Stelle in der Geschichte der modernen Selbstentfremdung genau zu bestimmen." (Heinemann 1954, 53)

[30]"Erst der Rückblick aus der nunmehr gewonnenen geschichtlichen Distanz läßt es verstehen, dass sich in diesem Text ein *Abschied von maßgeblichen Traditionen des neuzeitlichen Denkens und der Aufbruch auf einen neuen Boden denkender Besinnung* vollzieht." (Landgrebe 1962, 135)

[31]"Es ist ein widerwilliger Abschied, indem Husserl diese Tradition vollenden und erfüllen wollte, ohne dass es ihm sogleich bewusst wurde, wie sehr er sie bei diesem Versuche sprengte." (Landgrebe 1962, 135)

[32]"Wenn heute vor allem auf Grund der Wirkung des Werkes Heideggers vom ‚Ende der Metaphysik' wie mit Selbstverständlichkeit gesprochen wird, so verstehen wir den Sinn dieser Rede erst richtig, wenn wir genau verfolgen, wie in diesem Werk die Metaphysik sich sozusagen hinter dem Rücken Husserls verabschiedet." (Landgrebe 1962, 136)

phenomenology speak of experience? (2) How does phenomenology characterize the subject of this experience? (3) Finally, how does phenomenology describe the field of experience?[33]

In the present context, only brief answers to these questions are possible. So as to answer the first question, one needs to draw a distinction between *mundane experience* and the *reflective self-experience of the "I am"*; it is the latter, not the former, that defines the sense of *absolute experience*, which becomes the subject matter of phenomenological descriptions. In response to the second question, a distinction between the *mundane* and the *transcendental* ego is called for; it is the latter, not the former, that is identified as the subject of absolute experience. Landgrebe's response to the third question is of particular significance, and for two reasons. First, it is in this response that one is to seek evidence for Landgrebe's claim that *First Philosophy II* presents us with a "history of a shipwreck," in which the whole tradition of thought that determines modernity comes to a close. And secondly, Landgrebe's response brings to light the central role that the problematic of the horizon plays in Husserl's phenomenology.

On Landgrebe's view, it is the problematic of the double reduction and intentional implications that provides phenomenology with the needed resources to describe the field of experience. One can perform the reduction upon individual acts and show how the entire field of experience is entailed in these individual acts. That is, in each and every experience, the transcendental ego can be said to know its past and its future, because each actual consciousness already implies a representation (*Vergegenwärtigung*), which can be transformed into acts of remembrance, anticipation, or phantasy.[34] But if the entire field of experience is already implied in each act of consciousness, then one can claim that the consciousness of the "I am," besides entailing remembrance, anticipation, and phantasy, also implies the consciousness "this, my world, is" as well as the consciousness of other subjectivities.[35] Thus the result of this analysis is as follows: the field of transcendental experience is "indeed a field and its correlate is the world as it is intended by consciousness" (Landgrebe 1970, 280).[36]

Such a result, however, is nothing other than the realization that the problematic of the double reduction and of intentional implications leaves phenomenology not just with isolated acts, but rather with a field of experience, which is given in the form of a *horizon-consciousness*. It is thus the consciousness of horizons that allows Landgrebe to answer the third question indicated above. For this reason, Landgrebe identifies the analysis of horizon-consciousness as one of the most important discoveries that Husserl had made in *First Philosophy II*.[37] Yet on Landgrebe's view,

[33] See Landgrebe (1962, 146).

[34] See Landgrebe (1962), pp. 151–152.

[35] See Landgrebe (1962), pp. 154–156.

[36] "Daher zeigt sich als *Ergebnis dieser Überschau* über das transzendentale Erfahrungsfeld, dass es wirklich ein *Feld* und ihr Korrelat die *Welt* als vermeinte ist." (Landgrebe 1962, 155)

[37] As Landgrebe also remarks, Husserl himself had viewed the analysis of the horizon an essential advancement made in these lectures.

the groundbreaking nature of this discovery is of much greater significance than Husserl could have anticipated: "with this theory of the implication of the total-horizon within every single actual consciousness an essential advance has been made beyond the entire tradition of the modern theory of consciousness which permits us to understand the limits of this tradition's position and the aporiae which result from its limitations" (Landgrebe 1970, 281).[38]

What underlies this claim is the apprehension that the horizon-problematic cannot be restricted to questions that address the general structures of consciousness. Such a methodological procedure is an illegitimate abstraction. This problematic must also embrace *intersubjective* and *historical* horizons. But if so, then Husserl's analysis, since it is undertaken in twentieth-century Europe, must in effect *situate* its own goals within the horizon of science and philosophy. Even more, transcendental subjectivity, with which the apodictic ground of all knowledge is associated, must also be situated within the historical horizon. It thereby becomes clear how the horizon-problematic leads to a set of issues that Husserl raises explicitly in his late and unfinished *Crisis*. However, and this seems to be the central point of Landgrebe's critique, by situating its own project within the historical horizon, phenomenology loses the means to obtain apodictic evidence as *fundamentum absolutum et inconcussum*. Such is the case because, presumably, the horizon that encircles any feature of experience renders any claim to apodicticity premature.[39] The task of phenomenology can no longer be that of isolating a particular claim as the fundamental ground on which all science and philosophy must rest. The historical dimension that is irreducible from the horizon-problematic carries the task of situating each and every claim within an historical horizon, from within which philosophical tasks, motives, and claims derive their sense and significance. The irreducibility of the horizon signifies that the ground of evidence that underlies any claim itself lies within a horizon, and thus the question of its apodicticity can only be answered on the basis of the analysis of these horizons. As Landgrebe sees it, this consequence does not stem from a break with the central goals of Husserl's phenomenology. On the contrary, it is a "consequence of a program dedicated to an ultimate establishing of philosophical truth upon 'absolute experience.' The way leading to this consequence was first opened by the analysis of the consciousness of horizons..." (Landgrebe 1970, 283).[40]

---

[38]"In der Tat ist mit dieser Lehre von der Implikation des Gesamthorizontes in jedem einzelnen aktuellen Bewusstsein, sei es Aktbewußtsein oder Zustandsbewußtsein, ein wesentlicher Schritt über die gesamte Tradition der neuzeitlichen Bewußtseinstheorie hinaus gemacht, der diese in den Grenzen ihrer Fragestellung und den dadurch entstehenden Aporien verstehen lässt." (Landgrebe 1962, 156)

[39]"So stellt sich heraus, dass die ‚absolute Erfahrung', auf die sich alle Verantwortung und Rechtfertigung des Lebens gründet, eine *geschichtliche* Erfahrung ist [...] das Ideal der Apodiktizität als unerfüllbar erwiesen hat." (Landgrebe 1962, 59).

[40]"Sie ist *kein Bruch* mit Husserls früheren Ansätzen, sondern die *Konsequenz* aus dem Programm einer letzten Begründung philosophischer Wahrheit auf ‚absolute Erfahrung'. Erst die Analyse des Horizontbewußtseins [...] hat den Weg zu dieser Konsequenz eröffnet" (Landgrebe 1962, 157–158).

Such, then, is the evidence that Landgrebe provides to support his provocative claim that *First Philosophy II* marks a shipwreck of transcendental subjectivism, understood as nonhistorical *apriorism* and modern rationalism, a shipwreck that occurs in Husserl's works, yet behind Husserl's back. Although on many grounds, Landgrebe's analysis to this day retains its profound significance, nonetheless, as I have already indicated in the beginning of this section, Landgrebe's analysis is not without its problems. The operative distinction between the Cartesian and the non-Cartesian Husserl—a distinction, which to this day is often assumed to be self-evident and definitive—is the central reason why a reevaluation of Landgrebe's interpretation of *First Philosophy II* is called for.

What is the basis that allows one to speak of the Cartesian and the non-Cartesian Husserl? Landgrebe's answer points in the direction of the double reduction and, most significantly, the consciousness of horizons. But such evidence is highly suspect, if only because the problematic of the horizon does play a significant role in Husserl's so-called Cartesian works. As we saw in the earlier chapters of the present investigation, *Ideas I*, which Landgrebe identifies as the most forceful expression of the Cartesian stance, is precisely the work in which the problematic of the horizon makes its first explicit appearance. And just as the problematic of the horizon does not originate in *First Philosophy II*, neither does the problematic of the double reduction. We already encounter Husserl's analysis of this theme in *Grundprobleme der Phänomenologie*, a set of lectures that Husserl had delivered in 1910–1911 and which was one of the important sources for *Ideas I*.[41]

Of course, one could respond to such counter-evidence with a simple claim that in these early works we can only find a blueprint of the problematic of the horizon and of the double reduction. Such indeed was the conclusion that my earlier analysis of the horizon has led to. And yet, as Landgrebe himself admits repeatedly, the conclusions he draws from *First Philosophy II* are built on undeveloped tendencies, some of which have found expression in the *Crisis*, while others remained unarticulated in Husserl's phenomenology. One could therefore say that just as *First Philosophy II* prepares the way for a new way into phenomenology that is spelled out in the *Crisis*, so the "Cartesian Husserl" prepares the way for the new way to the reduction that is spelled out in *First Philosophy II*. In both cases, such a way is prepared by the problematic of the horizon. But if so, then one seems to be in full right to claim that the departure from Cartesianism, of which Landgrebe speaks, is to be found precisely in those works, which he has indirectly labeled as Cartesian. Yet when so much is said, it is no longer clear what the qualification of Cartesianism is meant to designate.

In fact, the notion of Cartesianism, as it is employed in phenomenological discussions, is highly equivocal and I would suggest that this equivocity is covered up by Landgrebe's analysis. On the one hand, this notion qualifies a particular way to the reduction that Husserl has spelled out in *Ideas I*. In this regard, a departure from Cartesianism is nothing other than the search for alternative ways that would lead to phenomenology. On the other hand, the notion of Cartesianism qualifies a particular

---

[41] See Hua XIII, 178.

philosophical style, which invites one to turn one's attention away from things and the world and search for apodictic certainty, as the foundation of all knowledge and truth, in subjectivity. In this regard, a departure from Cartesianism is an abandonment of the task of finding the apodictic ground that underlies all philosophy and science. Yet this second sense of Cartesianism is not derived from Husserl's phenomenology, for whom the apodictic grounding of all knowledge always remained the fundamental task of philosophy. It rather derives from Heidegger's critique of Descartes, which is spelled out in *Being and Time* as well as in a number of his other works. In the context of Husserl's phenomenology, one is fully justified to draw a distinction between the Cartesian and the non-Cartesian ways to the reduction *in the first sense, yet not in the second one.*

Yet in "Husserl's Departure from Cartesianism" such a distinction between two senses of Cartesianism is not drawn, and the absence of such a distinction leads to significant consequences. For instance, consider how Landgrebe defines the limits of the Cartesian way to the reduction. On Landgrebe's view, the "Cartesian way" establishes a reference to the field of experience, and yet, this field remains largely indeterminate. As Landgrebe puts it, the Cartesian way "does not grant us insight into everything that has been overthrown with the discarding of this belief, nor does it grant us insight into what still remains with this evidence of the 'I am…'" (Landgrebe 1970, 278).[42] Such a perspective comes very close to how Husserl himself has spoken of the limits of the "Cartesian way" to the reduction in §43 of the *Crisis*. However, there is a subtle difference between Husserl's and Landgrebe's accounts, and this difference is not without consequence. For Landgrebe, the "Cartesian way" leaves the ego empty of content and for this reason, it fails to give a promising account of the field of experience. For Husserl, the "Cartesian way" brings the ego into view as *apparently* empty of content and thus it makes it all-too-easy for the reader to misjudge "*at the very beginning*" what is to be included in and excluded from the field of absolute experience. Thus while on Husserl's view, the problems with the "Cartesian way" are *didactic*, on Landgrebe's view, the problems with this way are *methodological*. And thus, while for Husserl, the Cartesian and non-Cartesian ways to the reduction are *complementary*, for Landgrebe, they are *incompatible*. Yet the reasons for this incompatibility derive from the above-mentioned ambiguity that qualifies the notion of Cartesianism. Supposedly, while the Cartesian way to the reduction is motivated by the task of discovering the apodictic ground of all knowledge, the non-Cartesian way that is hinted at in *First Philosophy II* leads to the realization that the ideal of apodicticity is unrealizable. Without some violence to Husserl's text, such a thesis cannot be justified. The recognition that the horizon-problematic plays a significant role *Ideas I* and other works that precede the publication of *First Philosophy II* invites one to abandon such a rough-and-ready distinction between the Cartesian and the non-Cartesian Husserl.

[42] "Nicht aber ist auf ihm zu ersehen, was mit diesem Umsturz alles umgestürzt wurde und was mit dieser Evidenz noch verbleibt, das heißt, wie diese augenblickliche und ganz punktuelle Gewissheit meiner selbst als des undurchstreichbaren Ich bereits weitere in gleicher Weise gesicherte Evidenz in sich schließt, mit denen sich ein 'transzendentales Erfahrungsfeld' eröffnet." (Landgrebe 1962, 152)

Arguably, the motives that underlie the inquiry into "two Husserls" have their resources not so much in Husserl's own works, but rather in those of Heidegger. Such is the case not only because of the ambiguity that surrounds the notion of Cartesianism, which springs from Heidegger's critique of modernity, but also because the talk of "two Husserls" is too reminiscent of the infamous *Kehre* in Heidegger's works. Even though there are numerous transformation that *First Philosophy II* brings forth, these transformations should be evaluated in the context of numerous other transformations that take place in phenomenology, transformations that stem from the *Logical Investigations,* Husserl's lectures on time delivered in 1905, *Ideas I*, Husserl's analysis of passive synthesis, the problematic of the *Crisis*, and others.

As I have mentioned above, Landgrebe's discussion of "two Husserls" was motivated by the aim to offer a response to the hermeneutical and existentialist critiques of Husserl's phenomenology. For this reason, the notion of Cartesianism operative in his work derives not only from Husserl's phenomenology, but also from hermeneutical and existentialist literature. Yet the equivocation provoked by such a strategy leads to rather paradoxical results. Even if one were to say that Landgrebe succeeds in showing that *First Philosophy II* is a precursor to hermeneutical phenomenology and existentialism, nonetheless, such an outcome would comes at a hefty price. While Husserl's departure from Cartesianism might hold significance in the context of the prevalent philosophical themes of the day, the same cannot be said about the "Cartesian Husserl." Indirectly, one succumbs to the hermeneutical and existentialist critique of phenomenology without questioning its validity. I will question the validity of this critique in Chap. 9.

I believe Landgrebe is in full right when he highlights a particular set of issues that has preoccupied Husserl in *First Philosophy II* and which later grew to be central themes in hermeneutics and existentialism. However, half a century after the publication of Landgrebe's work, a number of the themes addressed in *First Philosophy II* allow one to see Husserl's phenomenology not only as a precursor to hermeneutics and existentialism, but also as a viable alternative to the manner in which a number of themes are treated in these philosophical traditions. In what follows, I will aim to show how the phenomenological analysis of the origins of the horizon can enrich alternative accounts of this theme. In this regard, *First Philosophy II* marks an unprecedented achievement, although this achievement has little to do with the problem of Cartesianism in existentialism and hermeneutics. It is the realization that the horizon is a *horizon of reference* and a *horizon of validity*—a realization that derives from *First Philosophy II*—that has paved the way for the phenomenological inquiry into the origins of the horizon.

References

Aguirre, Antonio. 1970. *Genetische Phänomenologie und Reduktion*. Der Haag: Martinus Nijhoff.

Heinemann, F.H. 1954. *Existentz-philosophie lebendig oder tot?* Stuttgart: Kohlhammer Verlag.

Husserl, Edmund. 1950. In *Ideen zu einer reinen Phänomenologie und phänomenologischen Philosophie. Erstes Buch: Allgemeine Einführung in die reine Phänomenologie*, ed. Walter Biemel. The Hague: Martinus Nijhoff.

Husserl, Edmund. 1959. In *Erste Philosophie. Zweiter Teil: Theorie der phänomenologischen Reduktion*, ed. Rudolf Böhm. The Hague: Martinus Nijhoff.
Husserl, Edmund. 1973. In *Zur Phänomenologie der Intersubjektivität. Texte aus dem Nachlass. Erster Teil. 1905–20*, ed. Iso Kern. The Hague: Martinus Nijhoff.
Husserl, Edmund. 1976. In *Ideen zu einer reinen Phänomenologie und phänomenologischen Philosophie. Erstes Buch: Allgemeine Einführung in die reine Phänomenologie*, ed. Karl Schuhmann. The Hague: Martinus Nijhoff.
Husserl, Edmund. 2006. In *Späte Texte über Zeitkonstitution (1929–1934). Die C-Manuskripte*, ed. Dieter Lohmar. Dordrecht: Springer.
Kern, Iso. 1962. Die drei Wege zur transzendentalphänomenlogischen Reduktion in der Philosophie Edmund Husserls. *Tijdschrift voor Filosofie* 24(1): 303–349.
Landgrebe, Ludwig. 1962. Husserls Abschied vom Cartesianismus. *Philosophische Rundschau* IX: 133–177.
Landgrebe, Ludwig. 1970. Husserl's departure from Cartesianism. In *The phenomenology of Husserl*, ed. R.O. Elveton. Chicago: Quandrangle Books.
Merleau-Ponty, Maurice. 1976. *Phenomenology of perception*. Trans. Colin Smith. Atlantic Highlands: The Humanities Press.

Chapter 8
The Horizon and the Origins of Sense-Formation

8.1 The Question of Origins in Phenomenology

While it is largely recognized that the question of origins is of central significance for phenomenology,[1] both the *meaning* as well as the *possibility* of this question remain enigmatic. On the one hand, the question appears to be clouded in methodological obscurity: it is by far not obvious if phenomenology does not trespass the limits of its methodological framework when it directs its attention to what it calls the origins of sense-formation. On the other hand, it is also puzzling what by posing this question phenomenology aims to accomplish: to uncover the concealed history that underlies given phenomena—what significance can such an accomplishment have for phenomena under scrutiny?[2] The analysis undertaken in Chap. 7 proves its significance precisely in light of these questions. On the basis of the foregoing analysis, I would like to suggest that *the phenomenological question of origins derives its sense and its methodological justification from the genetic notion of the horizon, interpreted both as a system of validity and a system of references.*

Let me qualify the phenomenological question of origins as a question that aims to trace the concealed sense-accomplishments which underlie the configurations of sense and which embrace each and every appearing phenomenon. This qualification already indicates that the question of origins emerges on the basis of those horizonal structures I have addressed in Chap. 7. That is, the concealed sense-accomplishments can become a phenomenological theme only if the sense of appearing phenomena is not exhausted by the present intentional rays but also entails the sedimented habitualities of consciousness. Moreover, as we have already seen, so as to be effective, the sedimented accomplishments of consciousness need not be thematic; rather, for

[1] For the strongest expression of such a view, see Fink (1966).

[2] As Husserl himself formulates it in *Ideas I*, "für die Frage nach Sinn und Wert unserer Erkenntnisse sind die Geschichten dieser Erbschaften ebenso gleichgültig, wie es für den Wertgehalt unseres Goldes die Geschichte der seinen ist." (Hua III/1, 53)

S. Geniusas, *The Origins of the Horizon in Husserl's Phenomenology*, Contributions to Phenomenology 67, DOI 10.1007/978-94-007-4644-2_8,

the most part, these accomplishments remain unquestioned and concealed. Such being the case, to question these hidden accomplishments means to inquire into that "a priori"[3] which forms the sense of appearing phenomena.

Yet to disclose what underlies the question of origins is not yet to clarify how the question itself can be either meaningful or even possible. I would like to suggest that *while the horizon as a system of validity clarifies the sense of the question of origins, the possibility of the question is secured by the horizon as a system of references.* So as to show that such is the case, I will start with the second claim.

The phenomenological thematization of the "hidden subjectivity" finds its justification in the notions of the double reduction and of intentional implications. These notions account for how the non-manifest dimensions of sense, despite the fact that they evade phenomenality, find their place within that sphere that phenomenology aims to thematize. More particularly, the concealed accomplishments of consciousness can be treated phenomenologically because they are *implicated* within the present appearance or lived-experience. Thus their concealment notwithstanding, the dimensions of sense, which stem from the habitualities of the ego, can be transformed into phenomenologically accessible themes.

Yet if to raise the question of origins is to follow those intentional threads, which the notion of intentional implications unfolds and secures, then the question of origins is nothing other than an investigation directed at the horizon of subjectivity. Such is the case because, as we have already seen, the horizon of subjectivity reveals itself as the implication of the whole life of consciousness within each lived-experience. Thus to claim that consciousness is horizonal is to suggest that consciousness is a store of accomplishments; and to raise the question of origins is to render the genetic make-up of these accomplishments thematic by revealing how each of them either presupposes, or is presupposed by the other ones. We thereby see how the question of origins finds its methodological justification within the genetic notion of the horizon, when the latter is interpreted as a system of intentional implications.

Yet the horizon is not only a system of references, but also a system of validity, and while the former accounts for the methodological possibility of the question of origins, the latter discloses the task that underlies such a regressive inquiry. So as to uncover the motivational framework that underlies the posing of the question of origins, we need to follow further what the notion of intentional implications entails. This notion indicates that the mere givenness of phenomena does not guarantee the givenness of their sense; that in truth, this sense continues to evade us even when the *epochē* and the reduction secures the givenness of phenomena in a phenomenologically justifiable way. But if phenomena are given in such a manner that their sense remains concealed throughout their manifestation, then such an "empty" appearing clarifies the goal that underlies the posing of the question of origins: the task is that of uncovering those dimensions of sense which pertain to appearing phenomena, yet in such a way, that to a direct glance they remain concealed.

[3] For the phenomenological sense of the notion of a priori and its bond to the horizon, see EU, 32.

Otherwise put, the motive that underlies the question of origins consists in the need to uncover those hidden dimensions of sense, which render the phenomenality of phenomena possible.

Thus phenomena, as given here and now, remain the fundamental theme of phenomenology, even when it raises the question of origins. The directedness to past accomplishments is a detour phenomenology takes so as to reveal the full sense of what presently confronts us. The very posing of the question of origins is driven by the realization that the sense of the presently given phenomena does not "stare us in the face" and that it can be uncovered by means of a detour into the "history" of subjectivity.

8.2 Jean-Luc Marion's Challenge

Thus phenomena given "here and now" lie both at the *beginning* and at the *end* of phenomenology's regressive procedure. They lie at the beginning as a *motive* for undertaking an inquiry into those dimensions of sense, which simultaneously belong to and remain concealed from the given phenomena. They lie at the end as the *telos* of the regressive inquiry: it is the sense of the presently given phenomena that needs to be unfolded and clarified. Hence the question of origins is to be seen as an indirect route that one takes by abandoning the presently given phenomena, yet only so as to return to them and obtain them with an enriched sense.

But if so, then one is in full right to raise the question of the radicality of such a methodological maneuver. When one takes as a point of departure the presently given configurations of sense, and when one knows from the start that one performs a detour only so as to return to these very same pregiven sense-configurations, then does one not prescribe to the analysis a particular direction, and thus a significant limitation? Namely, does one not from the very start secure a space which admits only of a *justification* of the given sense-configurations, a justification which can only consist in the enrichment of the pregiven validity? And does such a state of affairs not indicate that the phenomenality of origins has from the very start been curtailed for the sake of a pregiven function that these origins were compelled to perform?

Let me exemplify this concern by turning to the framework within which Husserl's analyses of the origins of sense-formation are most trenchant. I am referring to Husserl's transcendental aesthetics, whose analysis we find in such works as *Analyses Concerning Passive Synthesis* and *Experience and Judgment*. Here the point of departure is formal logic as it has been handed down to us in its traditional determination (e.g., EU, §2). The task of raising the question of origins in this context is that of clarifying the never-thematized presupposition, which consists in the pregivenness of the *Gegenstände-worüber*, i.e., of objects about which judgments are formed. Such a return to origins consists in deepening traditional logic by way of supplementing it with transcendental logic. The unique achievement of Husserl's conception of transcendental logic consists in the realization that judgment is not the exclusive framework of either evidence or truth, that in fact the origins of these two notions is to be found already within the prepredicative experience. Thus the broadening of traditionally determined formal logic leads to the disclosure of how objects and the world already possess a meaningful structure and their own unique kinds of evidence, both of which motivate the emergence of logical forms.

In the subsequent sections of this chapter, I will deepen this so far all-too-general determination of the specifically phenomenological notion of transcendental logic and clarify what role transcendental aesthetics plays within it. I introduce its general determination only so as to exemplify the above-mentioned problem that pertains to the manner in which phenomenology raises the question of origins. On the one hand, the manner in which phenomenology broadens the notion of evidence is indeed an unprecedented disclosure of the rudimentary dimensions of sense which already belong to prepredicative experience. Yet on the other hand, it should cause deep concern that these dimensions of sense are disclosed from the standpoint of predicative judgment and for the sake of enriching the sense of predicative judgment. Thus it remains unclear how one can from the very start rest assured that the dimensions of sense that belong to prepredicative experience are indeed such that, as §7 of *Experience and Judgment* suggests, they are the *dynamis* which directs itself to predicative judgment as its *entelecheia*. Husserl has not explicitly ruled out the possibility that there are prepredicative dimensions of sense, which neither lend themselves to logical treatment, nor terminate in logical judgment. Yet even though not ruled out, such a possibility has nonetheless been suppressed: within the operative conceptual framework, it does not hold a chance to be rendered actual.

It is not hard to see that the problem in question pertains not just to the manner in which phenomenology raises the question of origins, but, even more fundamentally, to the very notion of the horizon. Indeed, it is the notion of the horizon that proscribes the possibility of obtaining such origins of sense, which would not be teleologically tied to the present configurations of thinking. Since the origins in question are thematized through the notion of intentional implications, they cannot help but remain intentionally bound to the present manifestation of phenomena in question. Moreover, this bond is such that the more rudimentary levels of sense-formation are disclosed as partly empty in that they have the open future horizon that they strive to bring to fulfillment.[4] We thus see that the possibility of prepredicative experience, conceived in the absence of any intentional bond to predicative judgment, cannot be either phenomenologically disclosed or even justified because it contradicts that methodological strategy, which renders the question of origins possible.

Such a methodological standpoint has been rendered contestable in a number of post-Husserlian phenomenologies. First in Heidegger and later in a number of French phenomenologists inspired by him, we encounter alternative models to how

[4]"Auch hinsichtlich der schon wirklich gesehenen Seite ertönt ja der Ruf: Tritt näher und immer näher, sieh mich dann unter Änderung deiner Stellung, deiner Augenhaltung usw. fixierend an, du wirst an mir selbst noch vieles neu zu sehen bekommen, immer neue Partialfärbungen usw., vorhin unsichtige Strukturen des nur vordem unbestimmt allgemein gesehenen Holzes usw." (Hua XI, 7)

the question of origins is to be posed. A comparative analysis of all these models lies beyond the scope of this work. Yet let us ask: does the horizonal framework, as it determines the manner in which the question of origins can be posed, by itself indicate phenomenology's failure to obtain the necessary distance from within which it could thematize appearing phenomena? Nowhere else is such a critical view argued for more forcefully than in the works by Jean-Luc Marion. Let us therefore interrupt the present analysis of the horizon in Husserl's phenomenology with a brief detour to Marion's critique.

In *Being Given*, as well as in the *Reduction and Givenness*, Marion identifies Husserl's "principle of all principles" as the fundamental accomplishment of classical phenomenology. As Husserl formulates it, "No conceivable theory can make us err with respect to the *principle of all principles*: *that every originary presentive intuition is a legitimizing source of cognition*, that *everything originarily offered* to us *in 'intuition' is to be accepted simply as what it is presented as being*, but also *only within the limits in which it is presented there*" (*Ideas I*, 44).[5] According to Marion, the breakthrough that the principle of all principles announces is to be understood in terms of its opposition to the principle of sufficient reason (Leibniz) as well as the principle of possibility (Kant).[6] Leibniz and Kant in distinct ways restrict phenomenality within the boundaries of possibility, which they impose on phenomenality "from without." While in Kant this imposition is carried through from the perspective of the "formal conditions of experience" that belong to the *finite* mind, in Leibniz a similar kind of restriction of phenomenality is established from the perspective of the infinite, or indefinite mind. By contrast, the principle of all principles liberates phenomenality from its estrangement by announcing that intuition need not succumb to any external authority and that it itself can function as the ultimate source of right and justification. Yet according to Marion, Husserl himself does not fully appreciate the breakthrough that this fundamental principle announces, which, supposedly, can be witnessed by the fact that Husserl still subjects intuition to two conditions of possibility, "themselves non-intuitive yet nevertheless assigned to every phenomenon" (Marion 2002, 185). Marion identifies the *horizon* and the *transcendental I* with the conditions in question. It is the first of these conditions that should excite our special interest.

In what sense can the horizon be said to be imposed on phenomenality as an external limit? As the horizon always anticipates what is not yet seen, it transforms the unseen into the pre-seen, into a merely "belated visible," thereby erasing from it its "fundamentally irreducible novelty" (Marion 2002, 186). The horizon limits phenomenality by way of excluding the possibility of its manifestation in terms of its unforeseen character. Marion thus likens the horizon to a "visual prison," to a

---

[5]"Am *Prinzip aller Prinzipien*: daß *jede originär gebende Anschauung eine Rechtsquelle der Erkenntnis sei, daß alles*, was sich uns *in der 'Intuition' originär*, (sozusagen in seiner leibhaften Wirklichkeit) *darbietet, einfach hinzunehmen sei, als was es sich gibt*, aber auch *nur in den Schranken, in denen es sich da gibt*, kann uns keine erdenkliche Theorie irre machen." (Hua III/1, 51)

[6]See Jean-Luc Marion (2002), pp. 179–184.

"panopticon broadened to the dimensions of the world, a panorama without exterior, forbidding all genuinely new arising" (Marion 2002, 187). His analysis of the horizon is thus geared toward the recognition of the need to test a new possibility: "shouldn't we imagine freeing givenness from the prior limit of a horizon of phenomenality" (Marion 2002, 187)? "Could certain phenomena exceed their horizon" (Marion 2002, 209)?

The central task of Marion's phenomenology consists precisely in finding the means to answer both questions affirmatively. He thus draws a distinction between three original figures of phenomenality. First, Marion speaks of phenomena "poor in intuition" and identifies them with the objects given to formal intuition in mathematics and to categorial intuition in logic (Marion 2002, 222). He further identifies the "common law" phenomena as the second original figure of phenomenality, exemplifying them with phenomena given to physics and natural science (Marion 2002, 223). This whole analysis is meant to secure the specificity of the third group of phenomena, which Marion identifies as "saturated." Even more, Marion's analysis is geared toward the realization that phenomenality, as it falls within the first two types, is not original, but rather dervies from saturated phenomena. *Saturated phenomena constitute the paradigm of phenomenality*: "my entire project … aims to think the common-law phenomenon, and through it the poor phenomenon, on the basis of the paradigm of the saturated phenomenon, of which the former two offer only weakened variants, and from which they derive by progressive extenuation" (Marion 2002, 227).

What are saturated phenomena? The definitions Marion provides indicate their distance and resistance to what Husserl identifies as the horizonal framework of experience. Marion qualifies saturated phenomena as those appearances in which "intuition always submerges the expectation of the intention," in which "intuition sets forth a surplus that the concept cannot organize [and] that the intention cannot foresee" (Marion 2002, 225). In direct contrast to the anticipatory horizons of sense, Marion speaks of an "intuition that fore-comes" (Marion 2002, 225). In the most straightforward of ways, Marion announces that in the case of saturated phenomena, "the I of intentionality can neither constitute nor synthesize the intuition into an object defined by a horizon" (Marion 2002, 226). Having thereby "liberated" phenomenality from the horizons, Marion insists that saturated phenomena establish the truth of all phenomenality precisely because they mark the givenness from which appearances arise prior to their subjection to extraneous laws.[7]

One might, of course, wonder if the saturated phenomena of which Marion speaks truly have a place within our experience. Let us therefore turn to the first exemplification of phenomena under scrutiny that Marion provides in *Being Given*. The example in question is of special significance in that it brings us back to our central concern: is the horizonal framework of experience robust enough to

[7]In Husserl too, as we have seen, one can speak of still horizonless appearances. Is this what Marion means by saturated phenomena? Not at all! The horizonless phenomena of which Husserl speaks would not be saturated (as Marion understands this term), but rather just the opposite—they would be identified as impoverished phenomena.

thematize phenomenality? Marion provides us with a historical example. The first figure of saturated phenomena is an *event*:

> When the arising event is not limited to an instant, a place, or an empirical individual, but overflows these singularities and becomes epoch-making in time (delimits a homogeneous duration and imposes it as "a block"), covers a physical space such that no gaze encompasses it with one sweep and encompasses a population such that none of those who belong to it can take upon themselves an absolute or even privileged point of view, then it becomes a historical event. (Marion 2002, 228)

Such a characterization of an event insinuates that nobody is in the position to describe it exhaustively, which Marion takes to mean that nobody can constitute it as an object: "put trivially, nobody ever *saw* the battle of Waterloo" (Marion 2002, 228). In virtue of the excess of sense that refuses to be restricted to any perspective, no matter how privileged it be, historical events break free from the horizons, yet not "from without," but rather "from within." That is, the excess of sense in question generates an infinity of horizons: military, diplomatic, political, economic, ideological, etc. "The plurality of horizons practically forbids constituting the historical event into *one* object and demands substituting an endless hermeneutic in time" (Marion 2002, 229).

So as to appreciate the full significance that this exemplification of saturated phenomena implies in regard to Husserl's phenomenology, let us re-transcribe it from the framework of actual empirical history to that of "transcendental historicity." Could it not be so that the aim of genetic phenomenology to account for the sense of the present phenomena in virtue of their reference to concealed origins of sense cannot disclose the full phenomenality of the given phenomena precisely because this phenomenological strategy operates on the basis of horizonal consciousness? That is, could it not be so that the genuine sense of experience is too saturated with sense to be appropriated by subjectivity and retained within a horizonal framework? And should one not therefore say that, while Husserl's inquiry into the origins of sense can account for "poor phenomena" and "common-law phenomena," "saturated phenomena" indicate the limit of his phenomenological strategy?

Marion notwithstanding, I would like to suggest that these questions should be answered in the negative. For let us ask, how else is one to recognize saturated phenomena as saturated if not *in virtue of the horizon*?[8] To conceive of an event in terms of how Marion defines it, one needs to follow the guide of the horizon one inhabits so as to recognize the refusal of the event to be exhausted within the limited framework of a particular horizon. Is it not the horizonal reference that each of us in such circumstances draws to further accounts, explanations, narratives, and experiences that puts us in the position to recognize the inexhaustibility of the event? What sense would it make to speak of the saturated nature of the battle of Waterloo, if one were to cut off all its bonds to the limited horizons that characterize the perspectives of Fabrice, Flambeau, or Chateaubriand? The possibility to recognize

[8] Putting it otherwise, to identify saturated phenomena as originary, one must be able to draw a distinction between a saturated phenomenon and other types of phenomena, such as poor phenomena and common sense phenomena. Yet how is one to distinguish between them if not in virtue of the horizon? Does this distinction itself not presuppose the pregivenness of the horizon?

saturated phenomena as saturated cannot be established "on the other side" of the horizons, as Marion himself admits. It is rather the recognition of the *finitude*, *openness* and *blindness* of the horizons we inhabit that puts us in the position to face phenomena as saturated.

So let us ask: does Husserl's analysis of the horizon proscribe the possibility of such a recognition? Marion assumes such to be the case because his critique of the notion of the horizon in Husserl's phenomenology is guided by two presuppositions. First, Marion assumes that the horizons are fixed from the start as general frameworks of sense that are imposed on phenomenality "from without." Secondly, Marion also assumes that subjectivity is fully aware of the dimensions of sense that constitute particular horizons one inhabits. Yet neither of these presuppositions can be justified in the face of Husserl's analysis of the horizon. In the framework of Husserl's phenomenology, it hardly makes much sense to speak of an imposition of the horizons upon experience because the central task of Husserl's analysis into the origins of the horizons culminates in the realization that the horizons crystallize *in* experience. It also makes little sense to assume the fixed nature of the horizons because Husserl's analysis is geared toward the realization of the *malleable* nature of horizonality: as it will become particularly clear in the culminating sections of the third part of this work, each and every new experience brings about a transformation of the horizons of experience.

It is, however, the presupposition of the self-transparency of horizonal-consciousness that is exceptionally problematic in the present context. Only in virtue of such a presupposition can one assume that Husserl's analysis proscribes the possibility of recognizing phenomena as saturated with sense that exceeds the horizons. Yet let us ask, is it not precisely the recognition of the finitude, openness, and blindness of the horizons that calls for a regressive inquiry into their origins? If the horizons one inhabits did not bespeak self-opacity, what sense would it make to pose the question of their origins? So as to obtain a clearer sense of how the question of origins is thematized in Husserl's phenomenology, a look at the nature of *reflection* will prove helpful.

8.3 Reflection as Besinnung

The phenomenological investigations of the origins of sense-formation are indeed *reflective* investigations, yet not in the sense in which reflection has been rendered thematic in *Ideas I*. In this text, Husserl identifies the notion of *Reflexion* with an objectifying act through which lived-experience becomes subsumable and thus lends itself to a phenomenological treatment (Hua III/1, §77, §78). The need to secure the possibility of apodictic self-givenness of consciousness is the driving force behind Husserl's analysis of *Reflexion*. Being led by such a task, Husserl starts out with the recognition that there are different kinds of reflection and that it is phenomenologically indispensable to differentiate between them. Thus in §78 of *Ideas I* he declares that "the task is to distinguish the different 'reflections' and analyze them all in a systematic order" (*Ideas I*, 177).[9] The systematic order that Husserl provides is governed epistemologically. That is, one is to differentiate between different kinds of reflection by means of asking what kind of justification supports them. Of central importance for Husserl is the distinction between those reflections that have an *absolute* right and those that have a *relative* right. To the first group belong perceptual and "retentive" reflections, while reflection, built upon remembrance, falls within the second group (Hua III/1, 168–169). Such a hierarchical ordering of the different kinds of reflection remains determinative within static phenomenology. One can therefore say that for static phenomenology, much like for Descartes and Kant, the interest in reflection is the interest in establishing the possibility of the evident givenness of the entire privileged sphere of subjective being.

When Husserl discusses the origins of sense-formation, he indeed recognizes reflection as providing the only possible access to the theme in question, yet the notion he employs is not that of *Reflexion* but rather that of *Besinnung*. Here we touch upon a crucial, be it a subtle, distinction. One can intimate the difference between these terms by accentuating the temporal dimension of *Besinnung* in contrast to the spatial dimension of *Reflexion*. While immanent reflection of *Ideas I* aims to take hold of the entire domain of subjective being in one stroke, *Besinnung* instigates a patient return-inquiry (*Rückbesinnung*) into those strata of lived-experience that escape direct thematization. Thus Husserl no longer treats reflection as *Besinnung* within the systematic framework that governed over the different types of *Reflexion* in *Ideas I*. Rather, as A. Steinbock has previously suggested, *Besinnung*, in contrast to *Reflexion*, "takes the criteria of clarity and distinctness as its goal rather than as the starting point of reflection. *Besinnung* is a type of reflection that recognizes how implications of intentional life and forms of pregivenness are also constitutive of phenomena" (Steinbock 1995, 83). Instead of obtaining the phenomenologically legitimate domain of self-givenness once and for all, *Besinnung* is a regressive inquiry, which continuously enlarges the confines of the phenomenological domain by means of exposing the path traversed by consciousness in its emergence.

This enlargement of the confines of the phenomenological domain is nothing other than the progressive uncovering of the origins of sense-formation, which I have addressed in the last section. As Husserl puts it in *Formal and Transcendental Logic*, "*Sense-Investigation* [*Besinnung*] signifies nothing but the attempt actually to produce the sense 'itself,' which, in the mere meaning, is a meant, a presupposed sense …. Sense investigation, we could also say, is radically conceived *original sense-explication*, which converts, or at first strives to convert, the sense in the mode of an unclear meaning into the sense in the mode of full clarity or essential possibility"(FTL, 9).[10]

[9] "Die verschiedenen 'Reflexionen' zu unterscheiden und in systematischer Ordnung vollständig zu analysieren, ist darin die Aufgabe." (Hua III/1, 165)

[10] "Besinnung besagt nichts anderes als Versuch der wirklichen Herstellung des Sinnes 'selbst', der in der blossen Meinung gemeinter, vorausgesetzter ist … Besinnung, können wir auch sagen, ist, radikal verstanden, *ursprüngliche Sinnesauslegung*, die Sinn im Modus unklarer Meinung in Sinn im Modus der Klarheitsfülle oder Wesensmöglichkeit überführt und zunächst überzuführen strebt." (Hua XVII, 13)

Such a characterization ties *Besinnung* to the question of origins with an indissoluble bond and thereby brings us back to the unresolved problem, which was addressed in the last section, namely, to the question if and how the regressive procedure is capable of obtaining the critical distance from within which it could engage the presently appearing phenomena. Within the notion of *Besinnung* lies the possibility of answering this question because, as Husserl has it, "*radical* sense-investigation [*Besinnung*], as such, is at the same time criticism for the sake of original clarification" (FTL, 10).[11]

How are we to understand such a *critical* function of *Besinnung*? So as to obtain an answer, we have to turn once again to the notion of the horizon. As we have already seen, the horizonal appearance indicates that phenomena can be given in an empty manner, i.e., that they can be given in such a way that their genuine sense eludes us. Yet precisely because the concealed sense is sedimented within the horizonal framework, the possibility of its restoration remains open.[12] The uncovering of the concealed sense follows the path opened by intentional implications, *yet such a following is not to be understood as filling in a delineated direction that is from the very start determinate and articulate*. Nowhere is such a predelineation of sense to be found. Rather, "here original clarification means shaping the sense anew" (FTL, 10).[13] The clarification in question, as Husserl explains, is to be understood as a two-step procedure. First, it is a clearer determination of the vague and indeterminate predelineations of sense that lie concealed within the horizonal framework. Secondly, such a determination provides the ground on which one can test those configurations of sense within which the appearing phenomena are given to us; otherwise put, one can test those prejudices which are operative when one confronts appearing phenomena. This two-step procedure is what Husserl calls the critical function of *Besinnung*.[14]

Thus the critical function of the question of origins is to be understood as a *reestablishment* of the sense of phenomena that "here and now" confront us. Such a critical function is rendered possible by the fulfillment of the horizons of sense that at the start are given as empty. Such a reiterated fulfillment, carried through by following the thread of intentional implications, provides one with a standpoint from within which one can engage anew in the presently given sense-configurations.

We are now in the position to see the undisclosed presupposition that underlies the viewpoint, according to which genetic phenomenology does not have the means to acquire the critical distance so as to critically thematize appearing phenomena. To claim that such is the case is to assume that phenomena, as they are given "here and now" at the start of the genetic inquiry into the origins of sense-configuration,

---

[11] "Radikale Besinnung ist *eo ipso* Kritik, die ursprünglicher Klärung dient." (Hua XVII, 14)

[12] "Ist deren lebendige Evidenz verflossen, so verbleibt ihre habituelle Leistung, mit der Möglichkeit einer zunächst leeren Restitution, die dann in der Leergestalt die bestimmte Sinnesvorzeichnung enthält." (Hua XVII, 14)

[13] "Die Klärung hat den Charakter einer neuen Sinngestaltung." (Hua XVII, 14)

[14] See Hua XVII, 14.

are the same as those phenomena to which the inquiry into these origins leads. Yet phenomena given "here and now" at the start and at the end of such a genetic inquiry are decidedly different. *While the former are given in empty horizons, the latter show themselves in horizons fulfilled.* Fulfillment here does not simply indicate a fullness that replaces a former emptiness; rather, the fullness in question comes into necessary conflicts with numerous anticipations, from which empty horizons are inseparable. Due to such conflicts, the reestablished sense is decisively different from the "guiding" one.

The difference between the *guiding* and the *reestablished* sense indicates that the pregivenness of a present sense-configuration should not be seen as an unjustifiable limit, which would allegedly suppress the range of possibilities that pertain to the manner in which phenomenology engages in the question of origins. Rather, the pregivenness of phenomena is to be seen only as a preliminary notion (*Vorbegriff*)—and phenomenology is indeed in need of such a notion, for without it, it would fail to provide the question of origins with a *guiding sense*. To follow this direction is to fulfill the horizons of sense that initially are given as empty and thereby to obtain the presently given phenomena as fulfilled.

Let us turn to Husserl's analysis of the origins of predicative judgment in *Experience and Judgment* so as to see how such a restitution of origins is carried through. These analyses will corroborate the position I am here defending as well as help us determine more closely the genetic notion of the horizon.

8.4 Three Levels of Horizonality

The preceding sections of this chapter have established that the horizon (1) provides the methodological justification for inquiring into the origins of sense-formation, and that it (2) discloses the task this inquiry aims to accomplish. Yet what role does the *theme* of the horizon itself play within the phenomenological genealogy? Faced with this question, let us turn to §22 of *Experience and Judgment*. This section is of exemplary significance because it presents an outline of the structure that governs over the phenomenological analysis of prepredicative experience—an outline, which with full force reveals the central place reserve for the notion of the horizon. This section provides the basis to maintain that *the analysis of prepredicative experience is in truth nothing other than the disclosure of the horizon on three distinct levels of manifestation.*

It should not, however, be overlooked that the three levels of horizonality pertain to experience *in its unobstructed* form. By this I mean that Husserlian genealogy aims to reveal the structures of experience as they pertain to perception in its still *unmodalized* manifestation. This should not be misconceived as though the modalities of experience, such as negation, doubt, or possibility exclusively belong to predicative thinking. One of the central accomplishments of Husserl's transcendental logic lies precisely in the disclosure that modalities, which are commonly taken to belong exclusively to predicative thinking (as one learns early on, modalities are *modalities of judgment*) already emerge within the prepredicative dimensions of experience.[15] Yet even though modalization is already operative at the basic genealogical levels, it nonetheless remains possible to disclose the structures of experience in terms of its even more rudimentary manifestation. Such is the case because *modalization is already a modification*, which, as the latter term already indicates, is possible only on the ground of more basic levels of experience. As Husserl has it, "if it is granted in general that modalization takes place in the way we have described, namely, as the becoming-uncertain of the object in its being such and such, then it is also presupposed that a part at least of the process of contemplation of the object is also unobstructed" (EJ, 103).[16] More particularly, modalization emerges as a rupture of the expectations that belong to experience in its more primordial form—expectations, which pertain to the presupposed manner of the object's manifestation. Expectations in question are nothing else than the *horizons* that pertain to unobstructed experience. We thereby once again find ourselves in the position to admit that the horizons are *horizons of validity*. Moreover, we are also in the position to recognize that in its most primordial form, the horizons of validity manifest themselves as *horizons of unmodalized belief*. Thus *prepredicative experience, in its premodalized form, is already horizonal*, and it is such in terms of *three distinct levels of manifestation*.

The first level is that of *simple apprehension and contemplation* (*schlichte Erfassung und Betrachtung*), which Husserl characterizes as the *lowest level* of the objectifying activity. Let me exemplify this dimension of experience with the help of an image of standing at the shore of the ocean and noticing a ship in the distance. As the ship emerges from within the distant horizon and draws closer to me, its particular prominences (*Abgehobenheiten*) thrust themselves upon me thereby absorbing and redirecting my perceptual interest from the unity of the ship toward its particular qualities. Yet nothing stands in the way of my involuntary resistance to these perceptual tendencies and of my refusal to comply: as the ship draws nearer, I remain oriented solely to its indeterminate unity, and if I do so, my perception is nothing more than what Husserl calls *simple apprehension and contemplation*.

At first glance, such a basic type of perception appears too elementary to be called horizonal. Yet Husserl insists that "even the first apprehension and initial simple contemplation already has its horizons" (EJ, 104–105).[17] Experience is horizonal already at this rudimentary level because the objects to which it directs its interest appear as always already *typified* objects, i.e., they are objects that are from the very start *familiar*. Thus in the case of our example, the appearing object is early on conceived as an appearing ship; and even if one were to object that, if not under all, then at least under some circumstances, as the ship emerges in the horizon, its

[15] See in this regard EU, §21.

[16] "Soll es nämlich überhaupt dazu kommen, daß Modalisierung … eintritt, Ungewißwerden des Gegenstandes hinsichtlich seines So-oder-so-seins, so ist schon zumindest ein Stück ungehemmten Fortlaufens der Betrachtung des Gegenstandes vorausgesetzt." (EU, 112)

[17] "Die erste Erfassung und zunächst schlichte Betrachtung hat ja bereits ihre Horizonte." (EU, 114)

earliest appearance is all-too-vague and indeterminate to be called an appearance of a particular object that we can already name, it nonetheless remains the case that the object in question is recognized as *material*, as *perceptual*, as *visual*, thus as an object of a certain, be it a very general *type*. Thus the givenness of objectivity to simple apprehension is already a manifestation of objectivity, which from the very start falls under the guidance of a typifying consciousness. In this sense, simple apprehension is already a basic form of horizonal consciousness.

Let us turn back to our example. Under normal circumstances, as the ship draws closer to us, the perceptual interest is no longer satisfied with the contemplation of the unity of the object; while keeping the object as a whole in grasp as my main theme, I redirect my interest toward the object's particular qualities: the ship becomes recognized as a passenger ship, a schooner, or a small fisherman's boat, as being of a certain shape and color, etc. Such a perceptual reorientation is what Husserl calls the object's *explicative contemplation* (*explizierende Betrachtung des Gegenstandes*) and characterizes the latter as the second level of the unobstructed prepredicative experience. The transition from simple apprehension to explicative contemplation is the course perceptual interest takes as it follows the direction of the awakened expectations. To follow this direction is nothing other than to penetrate the object's *inner-horizon* under the guidance of the predileneated sense that the inner-horizon in its pregivenness announces. Or as Husserl has it, "*Explication is penetration of the inner-horizon of the object by the direction of perceptual interest*" (EJ, 105).[18] This is to be understood in the framework of unmodalized experience: the penetration of the object's inner-horizon is the *fulfillment* of the protentional expectations through which the object announces itself as the coming of the anticipated to its originary givenness. Thus it belongs to the originary givenness of the ship to disclose itself in its inner-horizon as a ship of this particular type, form, color, etc.

Yet the perceptual interest may remain unsatisfied with the disclosure of the given object in its inner-horizon. Besides providing the object with its *inner* determinations, I can just as well direct my attention toward the object's *relative* determinations. To do this is nothing other than to penetrate the object's *outer-horizon*, which is what Husserl calls the *third level* of prepredicative experience. To return to our example, I can characterize the ship as *smaller* than the ones I usually see, as being at a certain *distance* from the coastline, I can also notice the seagulls flying *above* it, or other ships *around* it, etc. Thus the awakening of the relative determinations of a particular object depends upon the co-givenness of other objects in perception's outer-horizon.

With these three levels of horizonality, we obtain the general framework of the Husserlian notion of prepredicative experience. To spell out what is entailed in these levels is to disclose the resources of prepredicative experience. One can therefore say that in the framework of the Husserlian investigations into the origins of sense-formation, horizonality is nothing less than the space of sense which Husserl's tran-

[18] "*Explikation ist ein Hineingehen der Richtung des Wahrnehmungsinteresses in den Innenhorizont des Gegenstandes.*" (EU, 115)

scendental aesthetics aims to thematize. The notion of the horizon and the theme of origins thereby prove to be inseparable: *a phenomenological inquiry into origins is only another expression for a phenomenological disclosure of the horizonality of experience.*

8.5 The Evidence of Experience

Thus the phenomenological investigations into the origins of sense-formation reveal that experience is from the very start horizonal. Yet what, let us ask, proscribes the possibility of an experience which would not succumb to horizonal structures? The analysis undertaken in the last section only shows that the possibility of a non-horizonal experience is excluded from the phenomenological framework. Whether this exclusion is legitimate or not—to this question we have so far not obtained an answer.

One can derive the reason why *experience (Erfahrung) is always already horizonal* from two theses that are central to genetic phenomenology: *(1) experience is always objective; (2) experience from the start grasps objects as worldly.* As my further remarks will make clear, to say that experience is always already imbedded in the world is to pinpoint its horizonal framework. Thus the second claim I am now proposing will allow us to broach anew the theme, which in the context of static phenomenology proved so elusive. I am alluding to the *world as horizon.* However, the elucidation of this theme rests upon a clarification of the first proposition. Let us therefore set aside the theme of the world-horizon until the final part of this work (Chaps. 10, 11, and 12) and direct our more immediate attention to the analysis of the first proposition.

The realization that experience is essentially objective is indispensable for understanding the relation between formal and transcendental logic in Husserl's phenomenology. In its first and broadest determination, Husserl qualifies experience as the *evidence* of individual objectivities: "the self-evidence of individual objects makes up the concept of experience in the broadest sense. Experience in the first and most pregnant sense is accordingly defined as a direct relation to the individual" (EJ, 27).[19] Such a qualification of experience is problematic, primarily because of the highly uncommon employment of the notion of *Evidenz.* Let us attend to it briefly, for it will disclose the significance of the phenomenological notion of experience for traditional logic in general, and for predicative judgment in particular.

We are accustomed to employ the notion of evidence exclusively in the sphere of judgments. The judgments we call evident are those that live up to the ideal of absolute,

[19] "*Evidenz von individuellen Gegenständen macht im weitesten Sinne den Begriff der Erfahrung aus.* Erfahrung im ersten und prägnantesten Sinne ist somit als direkte Beziehung auf Individuelles definiert." (EU, 21)

apodictically certain cognition. The minimal (and by no means sufficient) condition for the attainability of such judgments consists in the compliance with the rules of formal logic. Thus from the perspective of traditionally determined sphere of judgment, Husserl's employment of the notion of evidence appears surprisingly unrefined and therefore it is by far not clear how it can contribute to our understanding of either cognition or truth.

In direct contrast to this authorized view, Husserlian phenomenology broadens the notion of evidence, and not only so that it could, within the boundaries of sensuous experience, embrace the givenness of individual objectivities; even more generally, in this phenomenology evidence becomes equated with the givenness of *any type* of objectivity: "as '*self-evident*,' then, we designate *consciousness of any kind which is characterized relative to its object as self-giving this object in itself*, without asking whether this self-giving is adequate or not. By this, we deviate from the customary use of the term 'self-evidence,' which as a rule is employed in cases which, rigorously described, are those of adequate givenness, on the one hand, and of apodictic insight, on the other" (EJ, 20).[20]

Far from signaling Husserl's unrefined use of the notion of evidence, the equation of evidence with the givenness of any kind of objectivity announces the submission of evidence to a *critique* which remained missing within the traditional determination of logic. "Men believed that they knew in advance what self-evidence is. They believed that they could measure every other item of cognition against ideal, absolute, apodictically certain knowledge. They did not suspect that this ideal of knowledge … could for its part also require a justification and originary foundation" (EJ, 18).[21] Phenomenology's critical engagement in the question of evidence takes the form of an investigation into its origins and its genesis—an investigation which reveals that the traditional determination of judgment, and thus of evidence, is *one-sided* in that it masks the resources from which the need to determine judgment and evidence spring. Why one-sided?

Within the context of this broadened notion of evidence, Husserl insists that the evidence of experience, as the evidence of individual objectivities, stands out as genealogically primary. This means that besides being explicitly suppressed and dismissed by logic in its traditional determination, such a broad notion of evidence is also tacitly presupposed by it. Husserl discloses this concealed presupposition by thematizing the *interest* that guides the beginning logician.

---

[20] "*Als evident* bezeichnen wir somit *jederlei Bewußtsein, das hinsichtlich seines Gegenstandes als ihn selbst gebendes charakterisiert ist*, ohne Frage danach, ob diese Selbstgebung adäquat ist oder nicht. Dabei weichen wir von dem üblichen Gebrauche des Wortes Evidenz ab… *Jede Art von Gegenständen hat **ihre** Art der Selbstgebung = Evidenz.*" (EU, 12)

[21] "Von vornherein glaubte man zu wissen, was Evidenz ist, an einem Ideal absoluter, apodiktisch gewisser Erkenntnis glaubte man jede Erkenntnis messen zu können, und kam nicht auf den Gedanken, daß dieses Ideal der Erkenntnis und damit auch die Erkenntnisse des Logikers selbst, die doch diese Apodiktizität für sich in Anspruch nehmen, ihrerseits erst einer Rechtfertigung und Ursprungsbegründung bedürfen könnten." (EU, 10)

At first glance it seems that the interest in question consists in the differentiation of a variety of forms of judgment, a differentiation that culminates in the genesis of particular logical structures. Yet such a characterization, although up to a degree justifiable, is by far not sufficient. *Of central significance is the recognition that even before the interest in the logical arises, cognition has always already done its work.* The interest in the logical necessarily presupposes such a pregivenness of prelogical cognition for in its absence, logic would not be in possession of its *Gegenstand-worüber*. "The activity of judgment is considered by our regressive subjective inquiry as an activity which is at the service of the striving for knowledge. Knowledge of what? Speaking quite generally, knowledge of what-is, of the existent" (EJ, 19).[22]

Now curiously, this "prelogical" cognition already manifests a variety of "logical" dimensions, and it is precisely these dimensions that attract the interest of the beginning logician. Thus the forms of judgment that attract the logician's interest are already stamped as either leading toward genuine truth or merely to its semblance. We can thereby understand Husserl's position: "the interest which the logician has from the beginning in these structures is not merely an interest in just any formation having such and such a form but an interest in formations which claim to be the precipitate of cognitions" (EJ, 16).[23] Such a pregiven precipitate of cognition indicates that *the difference between judgments that culminate in genuine knowledge and those that constitute merely a semblance of truth precedes the emergence of logical reasoning*. One can even say that the pregivenness of this difference is the motivational ground that secures the stance of the beginning logician.

Yet of what import is this phenomenological discovery of the motivational framework from which logic emerges? First, its pregivenness indicates that as the beginning logician directs his interest to the purely formal and as he realizes that what he thereby obtains are merely the negative conditions of possible truth, he is also well aware that there are other conditions that need to be met if any proposed claim is to be seen as constituting genuine knowledge. Secondly, and more importantly, the beginning logician also realizes that within those other conditions lie the resources from which the evidence of judgments springs. These other conditions pertain to what Husserl calls the *subjective conditions of intuitability* (*die subjektive Charaktere der Einsichtigkeit*). Husserl maintains that these subjective conditions indicate the never-thematized two-sidedness of the logical problematic: "on the one side, we have the question of the constitution of forms and their laws and, on the other, that of the subjective conditions of the attainment of self-evidence" (EJ, 17).[24]

---

[22] "Das urteilende Tun kommt bei unserer subjektiven Rückfrage in Betracht als ein solches, das im Dienste des Strebens nach Erkenntnis steht. Erkenntnis wovon? Ganz allgemein gesprochen, Erkenntnis dessen was ist, des Seienden." (EU, 11)

[23] "Das Interesse, das der anfangende Logiker an diesen Gebilde hat, ist nicht bloßes Interesse an irgendwelchen Gebilden von bestimmter Form, sondern Interesse an Gebilden, die den Anspruch machen, Niederschlag von Erkenntnis zu sein." (EU, 7)

[24] "Einerseits die Frage nach den Formbildungen und ihren Gesetzlichkeiten, andererseits die nach den subjektiven Bedingungen der Erreichung der Evidenz." (EU, 9)

To disclose the subjective conditions of intuitability is nothing other than to describe the manner in which individual objectivities are given to us. The notion of the subjective conditions of intuitability thereby proves to be the evidence of experience of which I spoke at the beginning of this section. We thereby see what it means to call the evidence of experience the undisclosed presupposition of predicative judgment. The evidence of judgment is itself possible only on the ground of a more rudimentary evidence of the givenness of objectivities.

The difference between objective evidence and the evidence of judgment by itself does not secure the path that would lead the logician from the domain of predicative judgment to that of prepredicative experience. One is still in need of recognizing that (1) an empty judgment is an intentional modification of an evident judgment; further, that (2) immediate evidence is more rudimentary than the mediate one; and finally, that (3) the immediate judgments are judgments about the most basic objects (*letzte Gegenstände-worüber*), which Husserl calls the final substrates (*letzte Substraten*).[25] Following these procedures, we attain a clearer glimmering of prepredicative experience. Of course, one cannot rest content with a glimmering, no matter how one qualifies it. We have reached the level at which we can show that *the pregivenness of objectivities itself emerges from within the world-horizon*.

Yet before turning to the problematic of the world-horizon, I would like to return to the hermeneutical critique of Husserl's phenomenology. Although I have already touched upon this critique in the opening sections of Chap. 6, the conceptual framework did not permit me to address it in all the necessary details. Now that the horizon has been disclosed in its genetic phenomenality as the horizon of subjectivity, we are in a good position to return to this theme and address it anew.

References

Fink, Eugen. 1966. Die phänomenologische Philosophie Edmund Husserls in der gegenwärtigen Kritik. In *Studien zur Phänomenologie 1930–1939*. The Hague: Martinus Nijhoff.

Husserl, Edmund. 1966. *Analysen zur passiven Synthesis: Aus Vorlesungs- und Forschungsmanuskripten, 1918–1926*, ed. Fleischer Margot. The Hague: Martinus Nijhoff.

Husserl, Edmund. 1969. *Formal and transcendental logic*. Trans. Dorion Cairns. The Hague: Martinus Nijhoff.

Husserl, Edmund. 1973. *Experience and judgment: Investigations in a genealogy of logic*. Trans. James Churchill and Karl Ameriks. Evanston: Northwestern University Press.

Husserl, Edmund. 1974. In *Formale und transzendentale Logik. Versuch einer Kritik der logischen Vernunft*, ed. Janssen Paul. The Hague: Martinus Nijhoff.

Husserl, Edmund. 1976. In *Ideen zu einer reinen Phänomenologie und phänomenologischen Philosophie. Erstes Buch: Allgemeine Einführung in die reine Phänomenologie*, ed. Schuhmann Karl. The Hague: Martinus Nijhoff.

[25] For a detailed account and justification of these steps, see EU, pp. 14–21.

Husserl, Edmund. 1983. *Ideas pertaining to a pure phenomenology and to a phenomenological philosophy. First book: General introduction to a pure phenomenology.* Trans. F. Kersten. The Hague: Martinus Nijhoff.

Husserl, Edmund. 1999. In *Erfahrung und Urteil. Untersuchungen zur Genealogie der Logik*, ed. Landgrebe Ludwig. Hamburg: Felix Meiner Verlag.

Marion, Jean-Luc. 2002. *Being given. Toward a phenomenology of givenness.* Trans. Jeffrey Kosky. Stanford: Stanford University Press.

Steinbock, Anthony. 1995. *Home and beyond: Generative phenomenology after Husserl.* Evanston: Northwestern University Press.

Chapter 9
Husserl's Notion of the Primal Ego in Light of the Hermeneutical Critique

9.1 The Hermeneutical Critique of the Primal Ego

The last two chapters have shown how the question of the origins of the horizon is to be understood in the context of Husserlian phenomenology. So as to recognize the significance of this question and this procedure, we are in need of opening a dialogue between phenomenology and hermeneutics. Such is the case because in no other philosophical traditions is the horizon-problematic as significant as in phenomenology and hermeneutics. And yet, these philosophical traditions thematize the horizon in significantly different ways. As I have remarked in the Introduction to this work, the question of the origins of the horizon is central to Husserl's phenomenology and almost completely forgotten in post-Husserlian analyses of the horizon. The present chapter will substantiate this claim as well as offer new ways of opening a dialogue between phenomenology and hermeneutics.

According to the standard hermeneutical critique of phenomenology, the future of the phenomenological tradition largely depends on its capacity to purify itself of the numerous misconceptions that characterize classical phenomenology. As Paul Ricoeur, for instance, famously puts it, the history of phenomenology is the history of Husserlian heresies.[1] Such, undoubtedly, is the ambivalent approach to phenomenology taken by philosophical hermeneutics. The main representatives of this philosophical tradition assert the need to continue the phenomenological legacy, yet they also add: to remain true to the spirit of phenomenology, one needs to liberate oneself from the letter of Husserl's writings. The hermeneutical "purification" of phenomenology is meant to "place subjectivity back in the world"; the "hermeneutical turn in phenomenology" is meant to give rise to the recognition of the intersubjective, historical, and linguistic nature of subjectivity. These slogans suggest that

[1] "Si bien que la phénoménologie au sens large est la somme de l'oeuvre husserlienne et des hérésies issues de Husserl." (Ricoeur 1987, 9)

S. Geniusas, *The Origins of the Horizon in Husserl's Phenomenology*,
Contributions to Phenomenology 67, DOI 10.1007/978-94-007-4644-2_9,

Husserl's analysis of subjectivity reinforces an abstract conception of the ego, which is non-linguistic, pre-social, in short, a "worldless" subjectivity.

What motivates hermeneutics to take such an ambivalent stance in regard to Husserl's phenomenology? Ricoeur provides the following answer: on the one hand, hermeneutics aims to retain phenomenology's "generous effort to respect the diversity of appearing and to restore to each of its modes its quota of strangeness … of otherness" (Ricoeur 1967, 113). On the other hand, hermeneutics distances itself from phenomenology's idealistic interpretation of its own enterprise, from the effort to "reduce all otherness to the monadic life of the ego" (Ricoeur 1967, 113–114). The hermeneutical critique of Husserl's phenomenology is first and foremost directed against the idealistically interpreted subjectivity that we find so deeply entrenched in Husserl's program.

Such an ambivalent stance in regard to Husserl's phenomenology is characteristic not only of Ricoeur's thinking, but also of hermeneutics as a whole. Arguably, the origins of such an ambivalent stance lie in Heidegger's critique of Husserl. Heidegger's reading of Husserl's phenomenology in the *History of the Concept of Time*, while it avows the phenomenological significance of categorical intuition, is geared toward the realization that Husserl's phenomenology fails to raise the question of being.[2] By this Heidegger means that the notion of consciousness operative in Husserl's phenomenology is conceived in accordance with the dominant interpretation of being, viz., in accordance with being conceived as *Vorhandenheit*.[3]

Heidegger's critique of Husserl's project sets the tone to how philosophical hermeneutics has engaged in phenomenology ever since. As seen from hermeneutical perspective, Husserl's analysis of subjectivity is not just one of the latest, but also one of the most elaborate falsifications of subjectivity. Paradoxically, for Heidegger, Gadamer, and Ricoeur, Husserl's obsession with the question of subjectivity is exactly what leads phenomenology to a fabricated notion of subjectivity. Ironically, Husserl's constant effort to portray subjectivity as the ultimate origin of meaning and intelligibility is exactly what gives rise to a fictitious notion of subjectivity. The reason seems to be straightforward: between the transcendental primacy of subjectivity and the facticity of *Dasein* one must choose; one cannot have it both ways.

Heidegger's critique of Husserl's phenomenology primarily addressed Husserl's early works (first and foremost, the *Logical Investigations* as well as parts of *Ideas I*). Ricoeur's critique of Husserl reasserted Heidegger's central thesis by turning to the works Husserl had written in the middle period (first and foremost, *Ideas I* and *Cartesian Meditations*). Gadamer's critique, to which I will now turn, primarily addresses Husserl's late phenomenology (the *Crisis*).

Gadamer's approach to Husserl's late and unfinished *Crisis* is just as ambivalent as were Heidegger's and Ricoeur's interpretations of Husserl's earlier writings.

[2] See especially Heidegger (1979, §13).

[3] According to Heidegger's central argument that he presents in the main part of this text (as well as later in *Being and Time*), our understanding of the being of subjectivity remains distorted for as long as we do not clarify the differences between three fundamental modes of being, which Heidegger christens as *Vorhandenheit*, *Zuhandenheit*, and *Existenz*.

On the one hand, Gadamer openly admits that the centrality of historicity and the life-world problematic in the *Crisis* is a major impetus that underlies his own hermeneutics. On the other hand, Gadamer is highly critical of the fact that Husserl's analysis culminates in the realization that historicity and the life-world are constitutive accomplishments of the primal ego. Sociality, linguisticality, historicity, worldhood—for Gadamer, these themes are of primary significance; for Husserl, they appear to be phenomena of secondary importance, deriving their sense from transcendental subjectivity. As Gadamer sees it, Husserl fails to appreciate the full significance of the indicated themes because he never questions the possibility that they could disrupt the alleged primacy of subjectivity.

In *Truth and Method*, we find Gadamer's most extensive analysis of Husserl's phenomenology in the context of the historical preparations that pave the way to the theory of hermeneutic experience. Here it becomes patently clear that Gadamer's central critique of Husserl's phenomenology is directed at the notion of *constitution*. What Gadamer finds problematic is not so much a particular constitutive schema that distinguishes Husserl's phenomenology from alternative constitutive frameworks. Rather, it is the very insistence that phenomenology must clarify phenomena in terms of their constitution that places a significant limit upon Husserl's phenomenology. The limit in question is *twofold*.

First, as Gadamer sees it, the constitutive framework within which Husserl addresses each and every theme leads to a phenomenologically illegitimate privileging of *subjectivity*. According to Gadamer, that subjectivity, of which phenomenology speaks under the heading of the primal ego, does not find any confirmation in phenomenological experience. Yet clearly, if Gadamer's critique were only directed at subjectivity, then it would seem possible for hermeneutics to embrace Husserl's concrete analyses with only slight modifications. It would seem sufficient to give up this "worldless" subjectivity without introducing any fundamental changes in the descriptive analyses themselves. And yet, the relation between hermeneutics and phenomenology is not as close as this first critique suggests.

Thus secondly, as Gadamer sees it, the constitutive framework significantly curtails Husserl's concrete analyses. For instance, in the context of his discussion of the concept of life in Husserl's phenomenology, Gadamer writes: "we might ask whether ... the genuine content of the concept of life does not become alienated when it is articulated in terms of the epistemological schema: deriving it from the ultimate data of consciousness" (Gadamer 1989, 250).[4] This is, in fact, a claim that Gadamer is willing to repeat in regard to each and every phenomenological theme that he finds to be of central significance.[5] For Gadamer, the privilege of constitution

[4]"Man fragt sich, ob nicht [...] der eigentliche Inhalt des Lebensbegriffes durch das erkenntnistheoretische Schema einer solchen Ableitung aus letzten Bewußtseinsgegebenheiten überfremdet wird." (Gadamer 1990, 254)

[5]In his analysis, Gadamer addresses four central phenomenological concepts: intentionality, the horizon, the life-world, and life. In his analysis of each of these themes, Gadamer argues that the constitutive framework of phenomenology leads to their misconceptions.

is an *epistemological* privilege, and it is from this privilege that phenomenology needs to be freed. This privilege does not only give rise to an abstract notion of subjectivity; it also curtails the phenomenological descriptions by limiting them to merely egological analysis. Thus in the context of the historical preparation for the theory of hermeneutical experience, Husserl's analysis is superseded by those of Count Yorck and Heidegger.

It is not hard to see what this critique means for the horizon-problematic. Gadamer's critique of the constitutive framework places a demand on hermeneutics to abandon the inquiry into the origins of the horizon. Supposedly, this inquiry gives rise to an abstract notion of subjectivity as well as deforms the phenomenon of the horizon. It thereby also becomes clear why, in the context of Gadamer's critique of Husserl's phenomenology, the notion of the *primal ego* (*Ur-Ich*) occupies a central place. Supposedly, this notion is a prime example of how phenomenology distorts our understanding of subjectivity. What is more, such a supposedly distorted conception of subjectivity underlies the very problematic of the origins of the horizon.

Yet how exactly does Gadamer carry out his critique of the *primal ego*? This concept seems to be so foreign to hermeneutics that even though it constitutes the central target of Gadamer's critique, Gadamer never comes close to spelling it out. Everything seems to be only obvious: the primal ego is the "absolute irrelative to which all relativity is related"; it is an ego that supersedes "the basis of the pregiven world" (Gadamer 1989, 246).[6] Presumably, such a notion bespeaks a distorted conception of subjectivity, conceived in accordance with the model of being Heidegger has termed *Vorhandenheit*. But if so, then clearly, such a notion cannot find hermeneutical justification. Thus after touching on the question of the primal ego in a sole brief paragraph, Gadamer turns to that "element in Husserl's thinking, that constantly threatens to burst this framework asunder" (Gadamer 1989, 246).[7] Gadamer has in mind Husserl's notion of *life*.

Why is the notion of life irreconcilable with that of the primal ego? Presumably, one can conceive of life only insofar as it unfolds within the life-world, which is intersubjective, historical, and linguistic through and through. These qualifications set the concept of life apart from the notion of the primal ego, to which none of the mentioned qualifications seem to apply. This contrast between the concepts of life and the primal ego forces hermeneutics to once again reassert its ambivalent stance in regard to phenomenology. One can easily hear in this contrast the hermeneutical demand to "purify" phenomenology, i.e., to free phenomenology from those distortions that allegedly plague Husserl's original program.

Thus in the context of the hermeneutical critiques of Husserl's phenomenology, the figure of the primal ego occupies a prominent place: this notion bespeaks Husserl's commitment to a notion of subjectivity that precedes its being-among-others and its

---

[6]"Sie ist das Irrelative schlechthin, auf das alle Relativität, auch die des forschenden Ich, bezogen ist" (Gadamer 1990, 252).

[7]"Indessen gibt es schon bei Husserl ein Moment, das diesem Rahmen in der Tat ständig zu sprengen droht" (Gadamer 1990, 253).

being-in-the-world—and to this notion of subjectivity Husserl continued to cling to the very end of his days. As seen from the hermeneutical perspective, the primal ego is the final figure that announces Husserl's unyielding resistance to the "hermeneutical turn in phenomenology." What is more, this notion of the ego forces Husserl to thematize the problematic of the horizon in a similarly distorted way, viz., by raising the question of its origins.

9.2 The Hermeneutical Critique and the "Functional Interpretation" of the Primal Ego

Yet is the notion of the primal ego as irreconcilable with the life-world problematic as Gadamer takes it to be? Such appears to be the case for as long as the primal ego is understood in accordance with what Heidegger calls *Vorhandenheit*. But is this conceptual framework characteristic of Husserl's own project? There are good reasons to be doubtful. The hermeneutical critique appears dubious as soon as one takes into account that *The Crisis* is not the only text in which the notion of the primal ego is to be found. We also find this notion in the *Bernau Manuscripts* as well as in the *C-Manuscripts*. It is both curious and telling that the notion of the primal ego changes its precise meaning in these texts. With this in mind, Dieter Lohmar has recently argued against the Kantian readings of the primal ego, which would in one form or another identify the primal ego with Husserl's "discovery" of the unity of apperception.[8] As Lohmar has it, the primal ego is "not a hypostasized or fixed ego and it cannot be regarded as something like the Kantian 'I think.' The precise meaning of each primal ego can only be determined in the concrete context of research in respect to different deeper levels of constitution—be it low-level time-constitution, be it the constitution of the Other in primordial reduction etc." (Lohmar 2007, 17).

I will refer to the reading that Lohmar represents as the "functional reading." For the "functional reading," the primal ego is always an ego *in relation* to a particular level of constitution. This means that the concept of the primal ego does not foreclose, but rather opens the possibility to take the next step in the regressive inquiry that will deepen and enrich the phenomenological analysis. Paradoxically, it is always possible to uncover a more rudimentary primal ego behind a particular figure of the primal ego that emerges within a particular framework. But if so, then it makes little sense to interpret the primal ego as the ultimate objective presence that underlies Husserl's phenomenology.

Nonetheless, I would suggest that the "functional reading" of the primal ego answers the hermeneutical critique in part, but not in full. One could argue that the

[8]See Dieter Lohmar (2009). An English translation is also available: "A History of the Ego in Husserlian phenomenology. The 'Arch-ego' in Husserl's Late Manuscripts On Time and the *Crisis*." Hereafter all references to this text address the author's own English translation.

functionalizing of the primal ego only *seems* to invalidate the hermeneutical critique, while in fact it only corroborates it. Two responses could be made on hermeneutics' behalf. First, one could ask: what does it really mean that the concept of the primal ego changes its meaning in different conceptual frameworks? Does it not indicate Husserl's failure to discover the primal ego? And how could Husserl not fail to discover it, when—so the critique would run—there is no primal ego to be found? Secondly, one could also argue that the mere fact that in Husserl we find a plurality of primal egos does not necessitate one to interpret this concept in a functional way. After all, all these primal egos are still egos, which means that they must stand in some kind of a relation to each other. But once the relation between different primal egos is brought to light, does one not have to privilege one particular figure of the primal ego over others? But if so, then is one not left with the realization that, for lack of a better term, the "ultimate primal ego," as far as it is ultimate, must be conceived in terms of *Vorhandenheit*?

As far as I can see, the functional reading of the primal ego does not have the resources needed to answer these objections. The hermeneutical objection calls for an inquiry into the relation between the different figures of the primal ego that we find in Husserl's works. More particularly, the hermeneutical critique provides the incentive to enrich a functional reading of the primal ego with a historical interpretation. How exactly do the different figures of the primal ego relate to each other? Is the ultimate figure of the primal ego not to be understood in terms of *Vorhandenheit*? These are the guiding questions that will guide my subsequent analysis. I would like to show that a direct engagement in these questions can bring to light something of great importance. It can bring to light that what the hermeneutic demand to abandon the problematic of the primal ego suppresses is not so much a phenomenologically and hermeneutically illegitimate conception of subjectivity; rather, what it suppresses is a meaningful alternative to the hermeneutical analyses of subjectivity. The Husserlian alternative suggests that subjectivity can be best understood by directing one's attention back to the "transcendental past," i.e., to the concealed accomplishments of the *hidden* subjectivity. While the Husserlian perspective is indeed an alternative to the hermeneutical standpoint, these two positions are not irreconcilable. Thus in what follows, I would like to suggest that the whole problematic of the pregivenness of subjectivity can disclose the concealed historicity of subjectivity, which in significant ways can *enrich* the hermeneutical accounts of self-understanding.

9.3 The Primal Ego in the Crisis of European Sciences

In the *Crisis*, the notion of the primal ego emerges as the final figure of subjectivity that completes the path to the reduction through the life-world problematic. How does this path differ from the earlier ones Husserl had taken? In contrast to the Cartesian path, which leads to the transcendental ego in one leap (Hua VI, 157–58;

Crisis, 155), the genetic paths to the reduction, of which the "path through the life-world" is an instance, instigate a patient return-inquiry that brings to light the numerous levels of the transcendental subjectivity's constitutive accomplishments.

One should not overlook the remark Husserl makes at the outset of his analysis of the primal ego: the genetic paths to the reduction inevitably lead to *paradoxical enigmas*. The performance of the epochē overcomes the naturalistic naiveté, yet it inevitably leads to *transcendental naiveties*, which make themselves apparent through the emergences of transcendental paradoxes. The notion of the primal ego emerges out of one such paradox, viz., the paradox of subjectivity (Hua VI, §53). On the one hand, Husserl's foregoing analysis that paves the way to the paradox of subjectivity has established that the world itself is a constitutive accomplishment of transcendental intersubjectivity. Yet on the other hand, universal intersubjectivity seems to be nothing other than humanity, conceived as the integral part of the world. Hence, the paradox: "the subjective part of the world swallows up, so to speak, the whole world and thus itself too. What an absurdity!" (*Crisis*, 180).[9]

A closer look reveals that the paradox we face is in fact *twofold*. On the first level, it is a paradox of intersubjectivity: the world is the accomplishment of intersubjectivity, which itself seems to be part of the world. At a deeper level, it is a paradox of subjectivity, taken in its "unique solitude": transcendental intersubjectivity itself is an accomplishment of subjectivity, yet subjectivity at first glance appears to be only one member of the intersubjective community. Again, what an absurdity!

Just as the paradox of subjectivity emerges in a twofold form, so the resolution that Husserl offers is also twofold. The first step addresses the manner in which the world could be said to be the accomplishment of intersubjective constitution. The second step leads one to the realization that transcendental intersubjectivity is itself a constitutive accomplishment of the primal ego.

At the first of these levels, it seems obvious that if the world is understood as a constitutive accomplishment of intersubjectivity, then intersubjectivity cannot be taken to mean a plurality of human beings, understood in the natural-objective sense, as real entities in the world. Yet does this realization not disrupt the very sense of intersubjectivity? How am I, as the one who performs the *epochē*, to obtain a sense of intersubjectivity if not through the manner in which the others manifest themselves precisely as other human beings? Husserl is confident that a genetic path to the reduction can overcome this difficulty: the very fact that the others are given to me as phenomena means that they can be conceived not only as fellow human beings, but also as poles for transcendental regressive inquiries. Thus although concretely, each "I" entails all its accomplished acquisitions that include the world itself, following a regressive inquiry, I can place in brackets all the worldly accomplishments of the Other and reveal the Other as a transcendental subjectivity; all the

[9] "Der Subjektbestand der Welt verschlingt sozusagen die gesamte Welt and damit sich selbst. Welch ein Widersinn." (Hua VI, 183)

Others are then conceived as functioning ego-poles to which the phenomenon of the world is given and within which it undergoes its constitution.[10]

Thus each and every "I" turns out to be both a mundane and a transcendental subjectivity. One might wonder whether such a solution to the paradox of subjectivity at the first (intersubjective) level does not also provide the resources needed to answer the paradox at the second level. Would it not suffice to say that just as Others, so I, too, am both a mundane and a transcendental subjectivity? This is where the above-mentioned distinction between the naturalistic and the transcendental naiveté reveals its significance. Such a claim, although it arises out of the suspension of naturalistic naiveté, would nonetheless be naïve at the transcendental level. The conjecture that the distinction between transcendental and mundane intersubjectivity might answer the paradox of subjectivity is built on a peculiar forgetfulness. "We have forgotten ourselves, the philosophers .... I am the one who performs the epochē, and, even if there are others, and even if they practice the epochē in direct community with me, [they and] all other human beings with their entire act-life are included, for me, within my epochē" (*Crisis*, 184).[11] Thus the distinction between transcendental and mundane intersubjectivity must be deepened with the realization that *transcendental intersubjectivity is itself a constitutive accomplishment* of that subjectivity which, in a "unique sort of philosophical solitude," constitutes in itself another as Other.

This is the subjectivity that Husserl calls the *primal ego* (*Ur-Ich*). And he adds that the primal ego is to be called "I" only by equivocation, even though it is a *necessary* equivocation. This means that on the one hand, the primal ego is not an "I" insofar as an "I" is an ego among other egos, be they mundane or transcendental. Yet on the other hand, the primal ego is an "I" insofar as it is the phenomenological residuum that remains after the performance of the primordial reduction. Thus on the one hand, the primal ego is unique in that it is indeclinable (*Crisis*, 185); yet on the other hand, "every human being who carried out the epochē could certainly recognize his ultimate 'I'" (*Crisis*, 186).[12]

[10] "Concretely, each "I" is not merely an ego-pole but an "I" with all its accomplishments and accomplished acquisitions, including the world as existing and being-such. But in the epochē and in the pure focus upon the functioning ego-pole, and thence upon the concrete whole of life and of its intentional intermediary and final structures, it follow *eo ipso* that nothing human is to be found, neither soul nor psychic life nor real psychophysical human beings; all this belongs to the "phenomenon," to the world as constituted pole." (*Crisis*, 183) "Konkret ist jedes Ich nicht bloß Ichpol sondern Ich in allen seinen Leistungen und Leistungserwerben, mitgerechnet die als seiend und soseiend geltende Welt. Aber in der Epochē und im reinen Blick auf den fungierenden Ichpol und von da auf das konkrete Ganze des Lebens und seiner intentionalen Zwischen- und Endgebilde zeigt sich eo ipso nichts Menschliches, nicht Seele und Seelenleben, nicht reale psychophysische Menschen—all das gehört ins 'Phänomen', in die Welt als konstituierten Pol." (Hua VI, 187)

[11] "Unser naives Vorgehen war in der Tat nicht ganz korrekt, und zwar durch die Selbstvergessenheit unserer selbst, der Philosophierenden, oder deutlicher gesprochen: die Epochē vollziehe *ich*, und selbst wenn da mehrere sind, und sogar in aktueller Gemeinschaft mit mir de Epochē üben, so sind für mich in meiner Epochē alle anderen Menschen mit ihrem ganzen Aktleben in das Weltphänomen einbezogen, das in meiner Epochē ausschließlich das meine ist." (Hua VI, 187)

[12] "Wohl aber könnte jeder Mensch, der die Epochē vollziehen würde, sein letztes, in all seinem menschlichen Tun fungierendes Ich erkennen." (Hua VI, 190)

Thus the second resolution of the paradox brings to light a *threefold* manifestation of subjectivity. First, the ego can be conceived as a mundane ego—as a member of humanity. Secondly, the ego can also be conceived as a transcendental ego—as a member of the transcendental intersubjectivity that constitutes the world. Thirdly, the ego can also be conceived as a primal ego—as the constitutive origin of transcendental intersubjectivity. The emergence of the primal ego in the *Crisis* overcomes a transcendental naiveté that overlooks the constituted nature of transcendental intersubjectivity.

Husserl's concrete analysis of the primal ego in the *Crisis* is lamentably brief. Yet arguably, Husserl's detailed account is missing because the problematic that falls under the heading of the primal ego has already been treated quite extensively in his other works, most notably, in the Fifth Cartesian Meditation. As Husserl puts it in the *Crisis*, the emergence of the primal "I"

> shows how the always singular "I," in the original constituting life proceeding within it, constitutes a first sphere of objects, the "primordial" sphere; how it then, starting from this, in a motivated fashion, performs a constitutive accomplishment through which an intentional modification of itself and its primordiality achieves ontic validity under the title of "alien-perception," perception of others, of another "I" who is for himself an "I" as I am (*Crisis*, 185).[13]

It is not just that the context, within which the notion of the primal "I" emerges, is the very same as the one we find in the Fifth Cartesian Meditation. More importantly, Husserl's description of how the primal "I" solves the paradox of subjectivity in broad strokes reiterates the main line of argument that is spelled out in greater detail in the last Cartesian Meditation. One can therefore say that the "epochē within epochē" that we find in the *Crisis* is a deepening of the transcendental epochē with the *primordial reduction*.

9.4 The Primal Ego in the C-Manuscripts

Is one then to say that the primal I, conceived as the residue of the primordial reduction, is the ultimate figure of subjectivity that we find in Husserl's phenomenology? The hermeneutical critique of this notion assumes a positive answer to this question. And yet, the manner in which the notion of the primal ego emerges in the *C-Manuscripts* invites one to answer this question in the negative. Admittedly, it is only natural to expect that a specific notion would have the same meaning in different texts written by one and the same author. Yet in regard to the notion of the primal ego

[13]"Sie kann aufweisen, wie das immerfort einzige Ich in seinem originalen in ihm verlaufenden konstituierenden Leben eine erste Gegenstandssphäre, die 'primordiale,' konstituiert, wie es von da aus in motivierter Weise eine konstitutive Leistung vollzieht, durch die eine intentionale Modifikation seiner selbst und seiner Primordialität zur Seinsgeltung kommt unter dem Titel 'Fremderfahrung,' Wahrnehmung eines Anderen, eines anderen Ich, für sich selbst Ich wie ich selbst." (Hua VI, 189)

in Husserl's phenomenology, such an assumption would be too rushed and, in the final analysis, illegitimate.

As we saw, the primal ego in the *Crisis* is conceived as the phenomenological residuum left unaffected by the performance of the *primordial reduction.* This reduction, as Husserl remarks in the Fifth Cartesian Meditation, is the reduction to the living present (CM, 133), conceived as the sphere of ownness that transcendentally underlies the intersubjective levels of world-constitution. In the *C-Manuscripts*, one of Husserl's central goals is to clarify time-constitution on the basis of the primordial reduction, i.e., on the level of the living present. One could therefore expect that this clarification would generate a much richer account of the primal ego than the one we find in the *Crisis.* Yet the *C-Manuscripts* lead to a significantly different notion of the primal ego. These manuscripts lead to the realization that the primordial reduction, as the reduction to the living present, is still infected with a transcendental naiveté. The reduction to the living present leads back to the figure of the primal ego *as it already finds itself in the primordial world.* Precisely because the reduction to the living present does not problematize the given relation between the ego on the one hand, and the primordial world on the other hand, this reduction cannot be conceived as the ultimate one, and it calls for another "epochē within epochē."

When in a manuscript from 1931, published as Text Nr. 20 in Hua XXXIV, Husserl once again introduces the notion of the primal ego, this notion is not only significantly different from the one that we find in the *Crisis.* In a sense, the primal ego of the *C-Manuscripts* is set against the primal ego of the *Crisis*: the former notion emerges after the latter is put in parentheses. Let us look at this matter in some detail.

At first glance, it seems that in the living present a phenomenologist can find everything that phenomenology has to bring to light.[14] Yet is the living present, assumed as the ultimate event (*Urgeschehen*) of transcendental subjectivity, to be conceived as a temporal process? Initially it seems that this question is to be answered affirmatively. Yet a closer look reveals that the specific temporality of the living present is to be conceived as an accomplishment of subjectivity. And thus, one is invited to perform a new "epochē within the epochē": "the naively won transcendental I must be again subjected to a transcendental reduction" (Hua XXXIV, 300).[15] This new reduction is meant to bracket those undisclosed apperceptions, which the reduction to the living present has naively presupposed without clarifying them. More particularly, the new reduction is meant to clarify how subjectivity, at the level of the living present, can already find itself within the primordial world.

If that transcendental subjectivity, which already stands in relation to the world, is a constitutive formation, then it must be possible to place it within brackets. Where is it that this new reduction leads us? It leads us to the transcendental primal

[14]"In der lebendigen Gegenwart dahinströmend und in ihrem urphänomenalen Wandel mit den darin auftretenden Wiedererinnerungen, Identifizierungen, Unterscheideungen *liegt alles*[...]." (Hua XXXIV, 298)

[15]"Das naiv gewonnene transzendentale Ich muss selbst wieder einer transzendentalen Reduktion unterworfen werden." (Hua XXXIV, 300)

ego and its transcendental primal life (Hua XXXIV, 300). This transcendental primal ego is conceived as the ultimate source from within which the concrete transcendental ego obtains its transcendental temporality; this temporality is arranged according to the strict temporal order of the modalities of the present, past, and future.

Thus this deeper reduction that follows the reduction to the living present highlights the distinction between two fundamentally different notions of the ego. At the level of the living present, the ego is inseparable from its acts, possibilities, and acquired habitualities; it is an ego that has already constituted the world and that continues to constitute it. Once the living present is itself subjected to an epochē, the ego, as the primal ego, is conceived as the primal ground of temporalization, i.e., as the ego of the all-temporalizing life.

Thus paradoxically, the *C-Manuscripts* provide an account of the primal ego that is "more basic" than the primal ego of the *Crisis*. As Husserl puts it in the *C-Manuscripts*, "in a certain sense one can say: while *all time originates out of temporalization*, all temporalization originates out of the primal *temporalization*" (Hua XXXIV, 300).[16] While the primal ego of the *Crisis* leads back to temporalization that genetically precedes the intersubjective constitution of the world, the primal ego of the *C-Manuscripts* leads back to primal temporalization that genetically precedes what Husserl calls the "primordial domain," discovered through the primordial reduction. Curiously enough, Husserl is well aware of this when he "discovers" the primal ego in the *Crisis*. This raises serious concerns in regard to the hermeneutical critique of the primal ego. Yet before turning back to this critique, let us ask: is the notion of the primal ego that we find in the *C-Manuscripts* the most rudimentary notion of the ego that we find in Husserl's phenomenology?

9.5 The Primal Ego in the Bernau Manuscripts

Yet again, this question needs to be answered in the negative. Husserl's analysis in the *C-Manuscripts* presupposes the analysis of time-constitution that has been carried through in his still earlier works, collected in Hua X and Hua XXXIII. Moreover, the *C-Manuscripts* are built on the working hypothesis that Husserl's earlier analyses have already solved the problems of time-constitution at the deeper level of hyletic affection, retention and protention.[17] Thus when Husserl argues in the manuscript addressed above that the primal ego is more rudimentary than the living present, he is in fact drawing on these earlier analyses. Yet curiously enough, the notion of the primal ego that we find in these earlier works is not identical with the one we find in the *C-Manuscipts*.

[16] "In gewisser Weise kann man sagen: *Alle Zeit entspringt aus Zeitigung*, und alle Zeitigung entspringt aus einer Urzeitigung." (Hua XXXIV, 300)

[17] See in this regard Lohmar (2007, 12).

To the best of my knowledge, Text Nr. 15 of the *Bernau Manuscripts* marks the emergence of the notion of the primal ego in Husserl's phenomenology as a whole. This text was written in 1918, thus more than one decade before the composition of the *C-Manuscripts* and almost two decades earlier than the *Crisis*. Moreover, from a genetic point of view, *Bernau Manuscripts* thematize the primal ego at more primordial levels of constitution than the *C-Manuscripts* and the *Crisis*. Thus just as in the *Crisis*, so in the *C-Manuscripts* too, Husserl's choice of the term "the primal I" is not to be understood as the discovery of the most original figure of the ego in phenomenology.

In Texts Nr. 14 and 15 of the *Bernau Manuscripts*, Husserl hypothesizes the possibility of a reduction that would lead to pre-egoic dimensions of experience. At first glance, the reduction to original sensuality seems to accomplish this goal. The ego is from the start identified as the correlate of instincts, affections, and acts. The reduction to original sensuality abstracts from all these dimensions of experience, and thus it is only to be expected that it will lead to a pre-egoic level of experience. Husserl is thereby led to draw a distinction between three different levels of experience: (1) the pre-egoic level of pure passivity, (2) the (already egoic) level of affections and reactions, and finally, (3) the level of attentiveness (*Aufmerksamkeit*), the level of the *intellectus agens* (Hua XXXIII, 276).

Here the reduction to the pre-egoic original sensuality is conceived as the reduction to immanent temporality. Yet Husserl's analysis leads to the realization that original sensuality is not pre-egoic at all. At first glance, after the performance of this reduction, "it appears that we now have everything subjective." Yet a closer look reveals that "in a certain sense we do, and yet we don't. What we have is that which is temporal, yet not everything subjective is temporal" (Hua XXXIII, 277).[18] What is not temporal is precisely I myself, conceived as the identical center, the identical pole, correlated with the whole stream of experiences. Much like in the *Crisis* Husserl was led to claim that the primal ego is called "I" only by equivocation, so here also he claims that this rudimentary "I" should not be called "I," that, in fact, it should not be called at all and should remain nameless.[19]

This means that the reduction to the original sensuality is still not free from a certain transcendental naiveté, and for this reason, it must be coupled with the reduction to the original time-constituting consciousness. The second reduction is called for by the realization that *inwardness*, which is supposed to be the phenomenological residuum, is in fact an ambiguous notion. On the one hand, inwardness can stand for the immanent contents of experience; on the other hand, this notion can also refer to the constituting consciousness within which these contents are given (Hua XXXIII, 281).

---

[18]"Scheinbar 'haben' wir damit alles Subjektive—und in gewisser Weise 'haben' wir es—und doch wieder nicht; denn was wir haben, ist eben Seiendes, Zeitliches, und nicht alles Subjektive ist Zeitliches." (Hua XXXIII, 277)

[19]"Das Ich sollte eigentlich nicht das Ich heißen, und überhaupt nicht heißen, da es dann schon gegenständlich geworden ist. Es ist das Namenlose über allem Fassbaren, das über allem nicht Stehende, nicht Schwebende, nicht Seiende, sondern 'Fungierende', als fassend, als werdend usw." (Hua XXXIII, 277–278)

This ambiguity generates two significantly different notions of the ego. On the one hand, as far as inwardness stands for the immanent contents of consciousness, the ego is what is constituted in the stream of consciousness along with the constitution of particular objects of consciousness. This ego is an actual and potential object of consciousness. In this regard, the ego is temporal. On the other hand, as far as the ego is conceived as the ego of the time-constituting consciousness, the ego reveals itself as all-temporal. This all-temporal ego, conceived as the correlate of time-constituting consciousness, Husserl calls the primal ego.[20]

On the basis of the foregoing analysis, Husserl is led to draw a distinction between the arbitrary and necessary elements of inwardness. On the one hand, everything that belongs to inwardness, as far as this term covers all immanent contents of consciousness, is accidental and arbitrary. There is no necessity that guides over the fact that at a particular moment of experience, this particular content of consciousness emerges in the stream of consciousness. On the other hand, what belongs to inwardness, conceived in terms of time-constituting consciousness, is necessary. More precisely, what is necessary is the givenness of the primal ego as the correlate of the arbitrarily given contents of consciousness.

It is difficult to ignore the structural similarities between this primal ego and the pure ego of which Husserl had spoken in *Ideas I* (Hua III/1, §57). Yet there is an important difference: while in *Ideas I*, the pure ego is conceived as the necessary correlate of intentional acts, in the *Bernau Manuscripts*, the primal ego is conceived as the correlate of pre-intentional sensual primal-impressions (*sensuelle Urimpressionen*).

It is clear that what Husserl calls "the primal ego" in the *Bernau Manuscripts* is something quite different from what this term means in the *Crisis*. While in the latter text, the primal ego is conceived as the origin of intersubjectivity, in the *Bernau Manuscripts* it is conceived as the necessary structural correlate of sensuous primary impressions. Yet how exactly does the "primal ego" of the *Bernau Manuscripts* differ from the "primal ego" of the *C-Manuscripts*?

With this question in mind, let me once again return to the passage from the *C-Manuscripts* to which I have already referred above: while *all time originates out of temporalization*, all temporalization originates out of the primal *temporalization*" (Hua XXXIV, 300). This passage allows one to draw a distinction between the three figures of the primal ego I have discussed above. (1) In the *Crisis*, Husserl identifies the primal ego with the level of temporalization; this ego is primal in relation to "all

[20]Husserl himself provides a rather vivid example to illustrate this point. Consider realizing that your feet are cold. How does this experience unfold on the three levels of experience Husserl had distinguished in Text Nr. 14? For this experience to emerge at the highest level (that of the *intellectus agens*), I would have to turn to this experience as a theme of my consciousness. At the lower level, that of affections, I would be affected and irritated by the coldness of my feet while being thematically preoccupied with other themes (for instance, while reading a captivating novel, I could press my feet closer to the heater). Yet just as affection precedes attention, so the sensuous tendencies precede affection. Thus at the lowest level, that of original sensuality, I would still in a peculiar sense experience the growing power of affection which has not yet turned into affection.

time"—that level, within which transcendental intersubjectivity is constituted. (2) In the *C-Manuscripts*, Husserl identifies the primal ego with the level of primal temporalization; in this regard, the ego is more rudimentary than the domain discovered by the primordial reduction. (3) The *Bernau Manuscripts* mark the discovery of the primal ego as the correlate of primal temporalization. As such, this notion of the primal ego is more basic than the other two figures of the primal ego that we find in Husserl's later works.

9.6 A Response to the Hermeneutical Critique of the Primal Ego

Having sketched the problematic of the primal ego in Husserl's phenomenology, let us now turn back to the hermeneutical critique of Husserl's notion of the primal ego. As far as the *exact formulation* of this critique in *Truth and Method* is concerned, clearly, this critique is an instance of misplaced criticism. The very fact that there are "more basic" primal egos than the one Husserl addresses in the *Crisis* means that Husserl's "discovery" of the primal ego in the *Crisis* cannot be conceived as the most rudimentary determination of transcendental subjectivity. Moreover, if there are further-reaching accounts of the ego than the one we find in the *Crisis*, then clearly, this account cannot be conceived in terms of what Heidegger calls *Vorhandenheit*.

Nonetheless, even though Gadamer's critique of the primal ego misses its target, the mere fact that Husserl's analysis of the primal ego is much richer than Gadamer takes it to be does not by itself mean that the hermeneutical critique does not hold. This fact only calls hermeneutics to abandon the exact formulation of the critique, not the critique itself. Thus I have suggested earlier that there are two ways in which hermeneutics can respond to the "functional interpretation" of the primal ego. First, hermeneutics can interpret Husserl's uncertainty regarding which figure of the ego is to be identified as the primal ego as phenomenology's inevitable failure to find the primal ego itself. Secondly, from the hermeneutical perspective, Husserl's identification of numerous figures of the ego as the primal ego does not foreclose the question regarding which of the primal egos is the most fundamental one; and once a certain figure is identified as the fundamental one, the hermeneutical critique can be redirected to it.

Let me turn to the first of these points. Could one not interpret the plurality of the primal egos as phenomenology's failure to discover the primal ego itself? According to such a view, Husserl's analysis of the primal ego inevitably leads to the subsequent discovery of a more rudimentary primal ego, precisely because there is no primal ego to be found. Yet clearly, such a view is indefensible: the "history" of the primal ego in Husserl's phenomenology is not to be conceived in terms of a progressive deepening of this notion; on the contrary, the primal ego of the *C-Manuscripts* (1931) is not as far-reaching as that of the *Bernau Manuscripts* (1918), just as the primal ego of Husserl's last and unfinished *Crisis* emerges in those dimensions of experience, which presuppose the primal ego Husserl had addresses in the earlier works. This means that when Husserl addresses the problematic of the primal ego in the 1930s, he is well aware that the primal ego is not primal at all, in the sense in which hermeneutics understands this notion. And this, in its own turn, means that Husserl does not intend his notion of the primal ego to be conceived in the manner in which hermeneutics conceives it. Thus it makes little sense to accuse phenomenology for its apparent failure to discover the "ultimate" primal ego when such a discovery never was phenomenology's objective.

Nonetheless, there is still a second way in which Gadamer's critique can be reformulated. If the primal ego of the *Bernau Manuscripts* is more fundamental than the primal ego that we find in the *C-Manuscripts* and the *Crisis*, then could one not redirect the critique formulated in *Truth and Method* to the *Bernau Manuscripts*? Just because there are many figures of the primal ego in Husserl's phenomenology is not a sufficient reason to brush the hermeneutical critique aside. Could one not say that even if Gadamer's critique of the primal ego in the *Crisis* misses its target, such is the case not because the target is not to be found, but rather because the target lies elsewhere, viz., in the *Bernau Manuscripts*? Should the primal ego of these manuscripts, conceived as the most fundamental figure of the primal ego, not be understood in terms of *Vorhandenheit*?

Yet if such indeed were the case, why would Husserl in his later works use the notion of the primal ego to qualify a less rudimentary manifestation of the ego? Furthermore, is it not counterintuitive to expect that one can find out more about the primal ego of the *Bernau Manuscripts* by opposing it to the other figures of the primal ego than by linking it to them? Moreover, why would Husserl—the most prolific philosopher of the last century—dedicate so few pages to this supposedly ultimate figure of subjectivity?

In the face of this second reformulation of Gadamer's critique of the primal ego, it is hard to overestimate the significance of Husserl's remarks in the *Crisis* that precede his analysis of the primal ego. As Husserl has it, "by carrying out the *epochē* the phenomenologist by no means straightaway commands a horizon of obviously possible new projects" (*Crisis*, 180).[21] Phenomenology begins by excluding "any underlying ground" and thus "must create a new ground for itself through its own powers" (*Crisis*, 181).[22] And yet, sooner or later, the phenomenologist is bound to find out that the created ground is still shot through with naiveties of various sorts; sooner or later, the phenomenologist is bound to give up the acquired ground and create new ones. In short, *the overcoming of the naiveté of the natural attitude leads to transcendental naiveties*. Husserl goes as far as to suggest that the *fate* of

[21] "Zudem verfügt der Phänomenologe keineswegs durch den Vollzug der Epochē ohne weiteres über einen Horizont selbstverständlich möglicher neuer Vorhaben." (Hua VI, 183)

[22] "Sie [Philosophie] muß also zunächst bodenlos anfangen. Aber alsbald gewinnt sie die Möglichkeit, sich aus eigener Kraft selbst einen Boden zu schaffen, nämlich indem sie sich in originaler Selbstbesinnung der in ein Phänomen bzw. ein Universum von Phänomenen verwandelten naiven Welt bemächtigt." (Hua VI, 185)

phenomenology is "to become involved again and again in paradoxes, which arise out of uninvestigated and unnoticed horizons" (*ibid*).[23]

If such is indeed the *fate* of phenomenology, then it becomes understandable why the primal ego of the *Crisis* does not exclude the possibility that phenomenology can discover more basic figures of the primal ego—those we find in the *C-Manuscripts* and in the *Bernau Manuscripts*. Moreover, if "a phenomenologist does not straightaway command the horizon of all possible projects," then one can never know in advance where exactly new transcendental naiveties are to be found. The notion of the primal ego is not meant to foreclose the possibility of a renewed inquiry into the transcendental origins of subjectivity's life; on the contrary, it is meant to overcome a particular transcendental naiveté and thereby open the possibility of new regressive inquiries. Putting it bluntly, one can never know if a particular figure of the primal ego is the ultimate one. The fate of phenomenology, conceived as the discovery and overcoming of ever-new transcendental naiveties, calls for renewed inquiries into the transcendental origins of world-constitution. It is thus only to be expected that there are other, even more rudimentary figures of the primal ego to be found alongside the one Husserl had thematized in the *Bernau Manuscripts*.

9.7 The "Functional Reading" and the Historicity of Transcendental Subjectivity

Just as the functional interpretation of the primal ego provokes hermeneutics to reformulate its critique, so this reformulated critique calls for a modification of the functional reading. To be more precise, the hermeneutical critique calls one to supplement the functional reading with an interpretation that highlights subjectivity's *transcendental historicity*.

As we saw, in the wake of the functional reading of the primal ego, one can reformulate the hermeneutical critique by raising the question regarding how the different primal egos relate to each other. The functional reading lacks the resources needed to answer this question in a promising way. The functional reading leaves one with a plurality of the primal egos without clarifying how they relate to each other. As Lohmar puts it, "the primal ego does not refer to one single entity …. the notion of the primal ego does not denote one single kind of ego" (Lohmar 2007, 16).

To be sure, the functional reading is within its rights when it insists that Husserl's analyses of the primal ego in the *Bernau Manuscripts*, *C-Manuscripts*, and the *Crisis* do not refer *one single primal ego*. Yet the mere fact that the notion "primal ego" changes its meaning in these diverse contexts does not warrant the conclusion that these three figures of the primal ego do not refer to "one single kind of ego."

[23] "Ihr Schicksal […] ist ein immer wieder neues Hineingeraten in Paradoxien, die von unbefragt, ja unbemerkt gebliebenen Horizonten herstammen und als mitfungierende sich zunächst in Unverständlichkeiten melden." (Hua VI, 185)

The functional reading notwithstanding, the primal egos *do* refer to one single kind of ego, viz., to transcendental subjectivity. *The three figures of the primal ego differ from each other in that they thematize transcendental subjectivity at three different levels of its constitutive accomplishments.* To ignore this would mean to ignore much of the philosophical significance that falls under the problematic of the primal ego in Husserl's phenomenology. It would mean to ignore Husserl's unique contribution to the philosophical thematizations of subjectivity.

To put this differently: Husserl's analyses of the primal ego do not refer to one single entity called "primal ego," yet not because there are three such entities. Rather, *the primal ego is not an entity at all.* A particular figure of the primal ego refers to a particular level of constitution, which initially remains hidden from phenomenology due to a particular transcendental naiveté. Moreover, the fact that Husserl speaks of different primal egos means that there are many levels of constitution, which remain affected by transcendental naiveté. Even more: by reminding us that there very well might be other primal egos to be found—even more rudimentary ones than the ones he himself has uncovered—Husserl invites one to further pursue a regressive inquiry into transcendental subjectivity. Such an open-ended regressive inquiry, as it proceeds from one level of transcendental naiveté to a deeper level, does not only provide a solution to a particular problem, encountered in the carefully circumscribed context of a particular paradox. Arguably, all these "technical difficulties" are in fact of secondary importance. More significantly, as it takes each regressive step, phenomenology brings to light a previously hidden dimension of subjectivity's transcendental life, thereby enriching our understanding of the horizons of transcendental subjectivity. It is this willingness to take an ever further step into the hidden dimensions of subjectivity's transcendental accomplishments that makes phenomenology so well-suited to enrich our philosophical self-understanding.

Thus the multi-level problematic of the primal ego brings to light the *historicity* of transcendental subjectivity. The task of the horizonal investigations is none other than that of clarifying this historicity. We face here a unique transcendental historicity: a historicity composed of numerous constitutive levels, where each particular level is built on a previous accomplishment. It is highly intriguing that for Husserl, such a transcendental historicity does not have a clearly defined starting point, just as it does not have a clearly defined endpoint. This means that, the hermeneutical critique notwithstanding, Husserlian subjectivity is not, as William James would say, a hook from which the world dangles; it is rather this apparently endless historicity itself.

One can thus say that when hermeneutics rejects the problematic of the primal ego, it does not eradicate a conception of subjectivity, which supposedly lacks hermeneutical and phenomenological justification. Rather, it suppresses a regressive inquiry into the *formation of the horizons* (*Horizontbildung*), which in significant ways could complement the hermeneutical analysis of the *fusion of the horizons* (*Horizontverschmelzung*). After all, within hermeneutics itself, regressive analysis is not out of place. In fact, Gadamer's rejection of the problematic of the primal ego is meant to open the possibility for a regressive inquiry that would lead to the recognition of the *intersubjective* nature of subjectivity. Gadamer's critique of Husserl's phenomenology is meant to show that the horizons subjectivity inhabits have intersubjective origins of sense. From the phenomenological point of view, this is undoubtedly true, *as far as a particular level—that of mundane subjectivity—is concerned.* Yet the recognition of the intersubjective nature of horizons need not preclude a further regressive inquiry; and arguably, Gadamer's misrepresentation of the primal ego as the ultimate phenomenological basis does precisely that.[24]

For hermeneutics, my belonging to the intersubjective horizons of sense is the most basic fact, beyond which no meaningful analysis can unfold. Even more: for hermeneutics, any attempt to go beyond the primacy of intersubjectivity would amount to its denigration. For this very reason, the question of the origins of the horizon in this philosophical context is illegitimate. Allegedly, this question is based on a distorted notion of subjectivity. It is based on the notion of the primal ego, yet not as this notion is thematized by Husserl himself, but rather as it is presented in *Truth and Method.* Phenomenology, in a manner of speaking, lifts this hermeneutical ban when it claims that subjectivity must have freed the space for the Other within its own horizon, so that the horizons of Others could appropriate my own self-understanding. Phenomenology reveals that the formation of horizons (*Horizontbildung*) in a significant way underlies and makes possible the fusion of the horizons (*Horizontverschmelzung*).

It is not difficult to see what hermeneutics would gain by giving up its critique of the primal ego—a critique, which, as the foregoing analysis suggests, is unjustified. If self-understanding, as Gadamer often suggests, is the fundamental hermeneutical theme, then by bringing to light how subjectivity "opens the space" for intersubjective horizons of sense, phenomenology can in significant ways complement and enrich our self-understanding.

9.8 The Horizon as the Thematic Space of Genetic Inquiry and the Horizon as the World's Pregivenness

Let me briefly retrace the path I have followed in the second part of this study. Starting with Chap. 6, I aimed to disclose the horizon as the conceptual space, which legitimizes the phenomenological question of origins. Thus I began with the disclosure of what it means to characterize the horizon as the horizon of subjectivity.

[24]Husserl's analyses of the primal ego are not supposed to be understood in terms of "objective presence" (*Vorhandenheit*), as Gadamer, following Heidegger, suggests. They are rather to be understood in terms of an open-ended pregivenness (*Vorgegebenheit*). What Husserl's analyses bring to light is a transcendental historicity of subjectivity, conceived as a continuous formation of the hidden horizons of sense that continue to guide over the life of subjectivity. Due to these horizonal formations, subjectivity can inhabit intersubjective horizons of sense that play such a crucial role in hermeneutics.

I then proceeded to show how a new path to the reduction, spelled out in *First Philosophy II*, enriches the notion of the horizon in such a manner that it lends itself to a genetic inquiry. Thus in Chap. 7, I thematized the horizon as a system of *references* and a system of *validity*. Moreover, the realization that these two systems are inseparable from each other gave rise to the full-fledged notion of the horizon of subjectivity. Now it is this horizon of subjectivity that further proved to be the conceptual space within which the question of origins could be disclosed as phenomenologically legitimate. On this basis, we could see that the inquiry into the origins of sense-formation is nothing other than an investigation of the crystallization of the horizon of subjectivity. Thus as we saw in Chap. 8, it is the horizon that (1) legitimizes the question of origins in phenomenology and that (2) discloses the method employed when the question of origins is raised. It is the horizonality of experience that always remains the main focus whenever the question of the emergence of sense-formation is posed. Or better, and this is something that is often overlooked in post-Husserlian phenomenology: the question of origins is called for by the horizonality of experience itself.

Still in Chap. 8, my analysis of the origins of sense-formation culminated in the realization that the world-horizon itself is the ultimate origin from which all sense-formations spring. The horizon thereby proves to be not merely what legitimizes the question of origins; even more fundamentally, it reveals itself as the fundamental origin of sense itself. Having seen in Chap. 9 how the phenomenological inquiry into the origins of sense-formation can enrich the hermeneutical analyses of self-understanding, let us turn to Husserl's investigations of the world-horizon, which itself becomes a phenomenological theme only due to the horizonality of experience. It will soon become apparent that the historicity of subjectivity, which I have addressed in Chap. 9, cannot be understood without taking into account the problematic that falls under the title of the world-horizon.

References

Gadamer, Hans-Georg. 1989. *Truth and method*. Trans. Joel Weinsheimer. New York: Continuum.

Gadamer, Hans-Georg. 1990. *Wahrheit und Methode. Grundzüge einer philosophischen Hermeneutik*. Tübingen: J.C.B. Mohr (Paul Siebeck).

Heidegger, Martin. 1979. *Prolegomena zur Geschichte des Zeitbegriffs*. Frankfurt am Main: Vittorio Klostermann.

Husserl, Edmund. 1960. *Cartesian meditations*. Trans. Dorion Cairns. Dordrecht: Kluwer Academic Publishers.

Husserl, Edmund. 1970. *The crisis of European sciences and transcendental phenomenology. An introduction to phenomenological philosophy*. Trans. David Carr. Evanston: Northwestern University Press.

Husserl, Edmund. 1976. In *Die Krisis der europäischen Wissenschaften und die transzendentale Phänomenologie. Eine Einleitung in die phänomenologische Philosophie*, ed. Biemel Walter. The Hague: Martinus Nijhoff.

Husserl, Edmund. 2001. In *Die 'Bernauer Manuskripte' über das Zeitbewußtsein (1917–1918)*, ed. Bernet Rudolf and Lohmar Dieter. Dordrecht: Kluwer Academic Publishers.

Husserl, Edmund. 2002. In *Zur Phänomenologischen Reduktion. Texte aus dem Nachlass (1926–1935)*, ed. Luft Sebastian. Dordrecht: Kluwer Academic Publishers.
Lohmar, Dieter. 2007. A history of the ego. The 'Arch-Ego' in Husserl's late manuscripts on time and the *Crisis*. Unpublished.
Lohmar, Dieter. 2009. Eine Geschichte des Ich bei Husserl. Mit Bemerkungen zum Ur-Ich in Husserls späten Zeitmanuskripten. In *Das Selbst und sein Anderes. Festschrift für Klaus Erich Kaehler*, ed. Pfeifer Markus and Rapic Smail. München: Verlag Karl Alber Freiburg.
Ricoeur, Paul. 1967. *Husserl: An analysis of his phenomenology*. Evanston: Northwestern University Press.
Ricoeur, Paul. 1987. *A l'école de la phenomenology*. Paris: Vrin.

Part III
The World-Horizon as the *Wherefrom*, *Wherein*, and the *Whereto* of Experience

Chapter 10
The World-Horizon as the *Wherefrom* of Experience

10.1 A Threefold Determination of the World-Horizon

If one casts a brief glance at the central themes that preoccupied philosophy in the second half of the nineteenth century, one notices that the notion of the world is not to be found among them. One of the central accomplishments of Husserl's phenomenology lies in having revived the problematic of the world as a meaningful philosophical theme in the aftermath of positivism.[1] The absence of the world-problematic in the late 1800s can be explained by philosophy's straightforward acceptance of the scientific notion of the world as a basic fact. Husserl's phenomenology brought to light that the scientific notion could not be the ultimate one in that it inevitably presupposes a more rudimentary world-experience.

It is largely acknowledged that Husserl's most robust analysis of the world-problematic is to be found in his last and unfinished *Crisis*. Therefore, when the phenomenality of the world is addressed in the works dedicated to Husserl's phenomenology, the notion of the life-world is granted inexorable priority. This is done in a twofold way: either by limiting the analysis of the world to what is entailed in the notion of the life-world, or by providing a more systematic account which incorporates Husserl's analyses of the world-horizon in his other works, yet only so as to reveal their prefatory nature in relation to the life-world problematic.[2]

However, the central subject matter of this investigation is not the life-world but the world-horizon, and therefore, I cannot follow either of the delineated paths.

---

[1]See in this regard Ludwig Landgrebe's "Welt als phänomenologisches Problem" (in Landgrebe 1963). For a more elaborate exposition of this position, see also the Introduction and Chapter II in Landgrebe (1952).

[2]Gadamer's "Die phänomenologische Bewegung," "Wissenschaft von der Lebenswelt" and Paul Janssen's *Geschichte und Lebenswelt. Ein Beitrag zur Diskussion von Husserls Spätwerk* represent the first strategy, while a number of essays by Ludwig Landgrebe collected in *Phänomenologische Geschichte* as well as Karl Schuhmann's *Die Fundamentalbetrachtung der Phänomenologie: Zum Weltproblem in der Philosophie Edmund Husserls* exemplify the second strategy.

S. Geniusas, *The Origins of the Horizon in Husserl's Phenomenology*,
Contributions to Phenomenology 67, DOI 10.1007/978-94-007-4644-2_10,

Even though the problematic of the life-world and the world-horizon in many regards overlaps, there nonetheless remain significant differences between them. One can characterize the central difference between the notions in question by saying that the life-world is a particular figure of the world-horizon, i.e., it bespeaks one way in which Husserl has thematized the world-horizon. And even though this way in some respects is further-reaching than the other ones, it nonetheless neither exhausts nor clarifies the different dimensions of sense that are entailed in the notion of the world-horizon.

I will not, however, begin with Husserl's account of the life-world so as to reveal its limitations and to supplement its shortcomings with the resources his other texts provide. Rather, my task will be that of systematically disclosing the different senses that the notion of the world-horizon entails. It is my contention that Husserl himself has not distinguished the senses in question, and therefore his account of the world-horizon, as well as its relation to the life-world, remain ambiguous. My task, therefore, is that of systematically distinguishing the central and indispensable dimensions of the world-horizon.

Consider A VII 1, one of Husserl's manuscripts from 1933 to 1934, published in Hua XXXIX. At the beginning of this manuscript, Husserl asserts that the world as nature is always given with a core of direct and immediate intuitivity. A phrase like this one seems to imply that the same background/foreground schema we have already seen employed in Husserl's investigations into the horizons of concrete objectivities guides the analysis of the world-horizon. Yet a few pages later, Husserl adds a surprising footnote: "it is reflection, though, that makes the world thematic on the ground of the antecedent and unthematic world-horizon" (Hua XXXIX, 83).[3] This difference between the world's reflective and pre-reflective manifestation gives rise to an arresting distinction between world-experience (*Welterfahrung*) and world-consciousness (*Weltbewußtsein*). While world-experience is said to be thematic and up to a degree intuitive, world-consciousness is pre-thematic and pre-intuitive:

> One should not, however, overlook that the world is not constituted the way singular realities are constituted. The world is a horizon, which is originally and constantly self-changing and yet remaining unitary; it is the unthematic horizon of all singular realities. To make the world thematic and, in a certain way, to direct to it a thematic regard, to want to know the world "experientially," to want to bring the world to intuition as the universe of possible experience … the unthematic world-horizon precedes all of this. And so the problem of the horizon remains here the counterpart of the problem that falls under the title of world-experience. (Hua XXXIX, 83)[4]

---

[3] "Das ist aber Reflexion, welche Welt schon thematisch macht auf Grund des Welthorizontes, der als unthematisches Weltbewußtsein vorangeht." (Hua XXXIX, 83, Anm. 1)

[4] "Doch ist nicht zu übersehen: Welt ist nicht konstituiert wie einzelne Reale; sie ist ursprünglich ständig sich wandelnder und doch vereinheitlicht verbleibender Horizont in jeweilig einzelnen Realen unthematischer Horizont. Welt zum Thema machen und in gewisser Weise eine erfahrende Richtung auf Welt nehmen, Welt 'erfahrend' kennen lernen wollen, Welt als Universum möglicher Erfahrung sich zur Anschauung bringen wollen—all dem geht voran der unthematische Welthorizont. Und so bleibt hier das Horizontproblem gegenüber dem Problem, was unter dem Titel Welterfahrung vorliegt." (Hua XXXIX, 83)

Of central significance is Husserl's recognition that the world conceived as the totality of possible experience is itself already built upon a more rudimentary pregivenness of the world-horizon. What Husserl calls in this manuscript world-consciousness (*Weltbewußtsein*) is exactly what I would like to characterize as the world-horizon, conceived of as the *wherefrom* of experience. And as we will see below, the notion of world-experience (*Welterfahrung*) bespeaks the givenness of the world, conceived as the *wherein* and the *whereto* of experience. Keeping the distinction between *Weltbewußtsein* and *Welterfahrung* in mind, let us turn to the threefold sense of the world-horizon, beginning with its first determination.

10.2 Welthorizont, Welt-als-Horizont, and Horizont der Welt

In my attempt to bring to light the sense in which the notion of the horizon would capture the rudimentary manifestation of the world, let me begin with a terminological observation, which is by far not merely of philological significance. The notion Husserl employs to characterize the horizonality of the world is that of *Welthorizont*. This notion lends itself to a twofold grammatical interpretation. On the one hand, *Welthorizont* can be taken to mean a contracted genitive and thus it can be seen as synonymous with the more awkward expression *Horizont der Welt*. According to such a view, just as one speaks of the horizons of objects, so one can also speak of *Welthorizont*, or *Horizont der Welt*—the horizon of the world. Yet on the other hand, one can also, and in direct contrast to the manner in which we thematize the horizons of concrete objects, speak of *Welthorizont* as synonymous with *Welt als Horizont*—the world-horizon, or world-as-horizon. It is my contention that our understanding of what is meant by the world-horizon in Husserl's phenomenology remains imprecise and distorted for as long as we do not take into account that *Welthorizont is not a contracted genitive but rather a compound nominative*.

As my allusion to the horizons of concrete objectivities already shows, the expressions *Horizont der Welt* and *Welt als Horizont* are by far not identical. The difference stems from an equivocity that pertains to the first expression: the genitive construction leaves undecided if the horizons in question specify the unique manner of the manifestation of the world, or if they are only relative in regard to the world, which remains transcendent to its own horizonal appearance. Due to this equivocity, the expression *Horizont der Welt* invites one to misconstrue the phenomenological sense of *Welthorizont* by reducing its *unique* horizonality to yet another manifestation of the horizons that embrace concrete objectivities.

Husserl, however, has not always been explicitly aware of the distinction in question. Precisely therefore in *Ideas I* he could have so easily transformed the question of the world's pregivenness into that of the world conceived as the totality of objectivities: after all, aren't both notions of the world meaningful on the basis of their horizonal manifestation? Yet by making such a quick move, Husserl obfuscated the equivocity, which stems from the fact that to conceive of the world either as a pregiven ground or as the totality of objects is already to operate on the basis of a *decidedly different* notion of horizonality. Let us take a closer look at the difference in question.

To speak of the horizon of the world as the totality of objectivities is to employ the same constitutive schema that pertains to each and every objectivity. In my account of this structure in the first part of this work, I distinguished between the core appearance, given directly and immediately, and the apperceptive layers of sense that embrace the direct manifestation and transforms it into an a*ppearance of objectivity*. Analogically, to conceive of the world as the totality of objectivities is to suggest that the world also has its core of direct and immediate manifestation, which in its own turn is apperceived as the *appearance of the world*. In both cases, the horizon reveals itself as something *relative* and *preliminary*: what is not given to me in direct intuition, either is or can be intuited by others, and thus can also be directly and immediate given to me. Thereby the possibility is always left open that the apperceived dimensions of sense can be transformed into direct appearance.

There can be no question that such a notion of the world is operative in Husserl's works. Consider, for instance, Husserl's description of how the world as earth is constituted in a manuscript from 1934, to whose closer analysis I will turn below:

> The openness as horizonality that is not completely conceived, represented, but that is already implicitly formed. Openness of the countryside—knowing that I have finally arrived at the borders of Germany—then arriving at the French, Danish, etc., countryside. I have not paced off and become acquainted with what lies at the horizon, but I know that others have become acquainted with a part further on, then again others yet with another part—representation of a synthesis of actual experiential fields which mediately produces the representation of Germany, Germany within the boundaries of Europe, and gives rise to a representation of Europe itself, etc.—ultimately of the earth. ("FI," 117–118)[5]

I do not want either to challenge the presence of such a notion of the world in Husserl's phenomenology or to dismiss its phenomenological legitimacy. As we will see in the subsequent chapters of this work, Husserl has the means to distinguish between the horizonality of objects and the world-horizon even when both presuppose the background/foreground schema. Nonetheless, I do want to suggest that so as to fully appreciate what the notion of the world-horizon entails, one needs to take into account that the most rudimentary manifestation of the world rests upon an essentially different constitutive framework.

In direct contrast to the indicated notion of the world, whose constitution is analogous to the constitution of concrete objectivities, in his later works, and particularly in the *Crisis*, Husserl calls the being of the world *unique* ("seiend in einer Einzigkeit" [Hua VI, 146]). He further qualifies the world's uniqueness by saying that in regard to it both plurality and singularity are senseless. Husserl thereby is led

---

[5]"Die Offenheit als nicht vollkommen ausgedachte, vorstellig gemachte, aber implizit schon geformte Horizonthaftigkeit. Offenheit der Landschaft—Wissen, dass ich schließlich an Deutschlands Grenzen komme—dann kommt französische, dänische, etc. Landschaft. Ich habe, was im Horizont liegt, nicht abgeschritten und kennengelernt, aber ich weiß, Andere haben ein Stück weiter kennengelernt, dann wieder Andere noch ein Stück—Vorstellung einer Synthese der aktuellen Erfahrungsfelder, die mittelbar herstellbar die Vorstellung Deutschland, Deutschland im Rahmen von Europa und dieses selbst ergibt, usw.—schließlich die Erde." ("Umsturz," 307–308)

to assert an essential difference between the being of the world and the being of objects: "there exists a fundamental difference between the way we are conscious of the world and the way we are conscious of things or objects" (*Crisis*, 143).[6] These indications intimate a phenomenological notion of the world, which calls for a different constitutive schema than the one at work in the phenomenological account of the constitution of concrete objectivities.

It should not be ignored that the very notion of *Welthorizont*, interpreted as a contracted nominative, already insinuates that there must be an essential difference between the consciousness of the world and the consciousness of objects. In the most straightforward of ways, this notion announces that the world does not come to appearance though horizons, but that *it itself is a horizon*. Thus in contrast to the *Horizont der Welt*, the notion of *Welthorizont*, once seen as synonymous with *Welt-als-Horizont*, can in no way be conceived as *relative* or *preliminary*. Yet if the horizonality of the world is non-relative, then we cannot account for it in terms of the background/foreground schema. We must rather assert that the horizonality of the world is *unique* in that it does not have a core of direct immediate appearance: *it is a background without foreground, a halo without any kind of intuitive core*. In contrast to the constitutive framework that governs over the appearance of concrete objectivities, let us specify the distinctive manner of the world's manifestation in terms of its *non-objective, non-thematic*, and even *non-intuitive* givenness.

Yet if it is neither objective, nor thematic, nor even intuitive, then on what grounds is one to maintain that the world as horizon is nonetheless *given*? Furthermore, how is one to transform its alleged givenness into a phenomenologically accessible theme? Confronted with this problem in *Welt und Endlichkeit*, Eugen Fink was led to the conclusion that being non-thematic, the world also remains non-given; and thus he went on to assert that only if one trespasses the boundaries of Husserlian phenomenology, can one render the world a phenomenologically accessible theme:

> For Husserl, the world is not given, just as it is not given for Kant. Kant says, it is only an "Idea," so Husserl says, it is only a "horizon." The world is reinterpreted as a non-thematic intentional structure of experience: It is conceived as a halo that is always already sustained in advance and which "emptily" encircles the field of experience.... It is only astonishing that a philosophy, which starts with the proclamation to consider everything in an unbiased and presuppositionless way, takes the peculiar being of the world so little into consideration. (Fink 1990, 149)[7]

Yet does the recognition that the world-horizon is non-objective, non-thematic, and non-intuitive necessarily lead to such a disconcerting conclusion? Do these

---

[6]"Es besteht aber ein grundsätzlicher Unterschied in der Weise des Weltbewußtseins und des Dingbewußtseins[...]." (Hua VI, 146)

[7]"Welt gibt es für Husserl nicht an sich, so wenig wie für Kant. Sagt Kant, sie ist bloß "Idee", so Husserl: sie ist bloß "Horizont." Die Welt wird umgedeutet zu einer athematischen intentionalen Struktur der Erfahrung: sie wird genommen als ein immerzu vorweggehaltener Hof, der "leer" das Erfahrungsfeld umrandet.... Das Erstaunliche ist nur, daß eine Philosophie, die ansetzt mit der Proklamation, alles unvoreingenommen und vorurteilslos betrachten zu wollen, so wenig das eigenständige Wesen der Welt in den Griff bekommt." (Fink 1990, 149)

characterizations necessarily mean that, as Fink further observes, the very mentioning of the world in Husserl's phenomenology is justifiable only because it depicts the standpoint from which Heidegger takes his departure?[8] I would like to suggest that such is not the case in the least. Despite its threefold negative qualification, the world remains not only a legitimate, but also one of the central themes of Husserlian phenomenology. So as to account for its legitimacy, we need to see how phenomenology obtains the means to describe what at first glance seems to lie outside the boundaries of phenomenological description.

In the *Crisis*, after indicating the difference between the way we are conscious of the world and of things, Husserl proclaims that these two types of consciousness nonetheless form an inseparable unity: 'each one [thing, object] is something, 'something of' the world of which we are constantly conscious as a horizon. On the other hand, we are conscious of this horizon only as a horizon for existing objects; without particular objects of consciousness it cannot be actual' (*Crisis,* 143).[9]

This necessary intertwining of the world-horizon and existing objects indicates a path that holds the promise of delivering us from the dead-end into which the non-phenomenality of the world-horizon seemingly has placed us. If the world as horizon is a horizon *for existent objects*, then the phenomenological description of the manner of givenness of existent objects must disclose how they necessarily emerge from within a more rudimentary pregivenness of the world. Otherwise put, *the world as horizon can be rendered thematic by way of disclosing how it forms a necessary counterpart to the appearance of existent objects so that without it, the appearance of objects would not be possible.*

Such a mutual overlap of the givenness of the world and of existent objects enriches the notion of intentionality. It indicates a *necessary two-sidedness that pertains to intentionality*: act-intentionality always goes hand-in-hand with horizon-intentionality as its necessary counterpart. Horizon-intentionality here must be understood not merely as a consciousness directed toward the horizons of objects but rather as a consciousness due to which subjectivity has a world. A closer determination of what the intertwining of the two types of intentionality amounts to will allow us to obtain a clearer sense of what it means to call the world a horizon.

10.3 World-Horizon as the Ground of Experience

My employment of the notion of ground in the framework of the horizon-problematic should not be misconceived as a suggestion, according to which the world would form a set of reasons or a causal nexus that would logically underlie the

[8] "Aber wir haben diese phänomenologische Auffassung kurz erwähnt, weil sie in einer gewissen Hinsicht den Ausgangspunkt Heideggers mitbestimmt." (Fink 1990, 149)

[9] "Jedes ist etwas, 'etwas aus' der Welt, der uns ständig als Horizont bewussten. Dieser Horizont ist andererseits nur als Horizont für seiende Objekte bewußt und kann ohne sonderbewußte Objekte nicht aktuell sein." (Hua VI, 146)

givenness of concrete objectivities. From Husserl's works we can extract two notions that lend themselves to be translated with one and the same English word "ground." The notions in question are those of *Grund* and *Boden*. While the first notion invites the above-mentioned interpretation, the second one elicits a different sense that just as strongly belongs to the word "ground": *Boden* is ground in the sense of land, soil, or simply ground beneath our feet.[10] We find an unequivocal distinction between these two notions in the *Crisis*, where Husserl employs an antonym of "Grund," viz. *Untergrund*, as synonymous with *Boden* (Hua VI, 127).[11]

These references to the *Crisis*, both in this and in the foregoing section, might give rise to the impression that the world-horizon I want to identify as the ground of experience is nothing other than the *life-world*. My intent, however, is quite different, and arguably, more radical. The world as *ground* of experience bespeaks a unique dimension of sense, which largely remains extraneous to the life-world problematic. The distinction I introduce between the world as ground and the life-world appears controversial given Husserl's frequent identification of the life-world as the non-thematic *ground*, which remains the most concealed presupposition of both scientific and philosophical reasoning. Nonetheless, one needs to assert that the life-world itself already rests upon a more rudimentary ground of experience, no matter how bewildering such a claim might sound. The distinction between *Weltbewußsein* and *Welterfahrung* necessitates that one establish such a distinction.

Following Ulrich Claesges,[12] let us distinguish three senses that the notion of the life-world entails. (1) The life-world in its narrow sense is to be understood by means of a contrast with objective sciences and the notion of the word that they give rise to: the life-world is the pre-scientific world. (2) The life-world in its broader sense is to be seen as a particular practical world, established through the leading idea that unifies and guides particular practices within the life-world. In this regard, the world of the objective sciences can be seen as a particular practical world. (3) The life-world in its broadest sense is the world taken in its full concretion, and as such it embraces both the first and the second determination of the life-world. I will turn to a closer analysis of the different senses that the life-world entails in Sect. 12.4. For now we only need to keep this equivocity in mind in regard to Husserl's identification of the life-world as the forgotten ground of the sciences. What is meant by such an assertion refers to the *first* notion of the life-world.

Yet the first determination clearly belongs to the domain of *Welterfahrung* and not to *Weltbewußtsein*. It is, as Husserl has it, a subjective-relative world. Thus it might well be non-thematic and non-objective, yet it nonetheless remains an *intuitive* world. But if so, then the life-world, even in its most narrow determination, must already presuppose a more basic *Weltbewußtsein* and thus rest upon a more rudimentary pregivenness of the world. It is this more elementary sense of the world's pregivenness that one needs to identify as the ground of experience.

[10] For a good illustration of the difference between these terms, see Steinbock (1995, 110).

[11] See Manfred Sommer's "Husserl on 'Ground' and 'Underground'" for a detailed account of how the notion *Untergrund* is employed in Husserl's phenomenology.

[12] See Claesges (1972), pp. 85–101.

A number of manuscripts that belong to the A VII group[13] suggest that the problem of the world-horizon at its rudimentary level must be seen as the problem of the determination of *Weltbewußtsein*. Yet how is one to understand the claim that everything thematic, intuitive, and objective rests upon and springs from a more primordial pre-givenness of *Weltbewußtsein*? With this question in mind, let us briefly turn back to *Ideas I*, to a passage already discussed on a few occasions.

"The seizing-upon is a singling out and seizing" (*Ideas I*, 70),[14] writes Husserl, and thereby emphasizes that the stream of experience cannot consist merely of what is *actually* given to consciousness. For Husserl, this claim bespeaks the primacy of potentiality over actuality, by which we are to understand that an actual appearance of objectivity always already rests upon its more rudimentary horizonal manifestation.[15] What we face here is a good illustration of the general manner in which Husserl interprets the intertwining of act-intentionality and horizon-intentionality in *Ideas I*: at issue is not the pregivenness of the world, but only the givenness of concrete objectivities.

It seems that Husserl repeats this claim verbatim, when in *Experience and Judgment* he writes: "but always preliminary to this grasping is affection, which is not the affecting of an isolated, particular individual" (EJ, 29–30).[16] Yet despite the patent affinities, a closer look reveals that the two claims are by far not identical. The difference between them is in fact crucial, for it hinges upon decidedly different notions of horizon-intentionality. While in *Ideas I* horizon-intentionality accounts for the non-thematic dimensions of sense that pertain to the givenness of concrete objectivities, in *Experience and Judgment* it directly bespeaks the pregivenness of the world:

> To "affect" means to stand out from the environment, which is always copresent ... The environment is copresent as a *domain of what is pregiven*, of a *passive* pregivenness, i.e. of what is always already there without any attention of a grasping regard, without any awakening of interest. We can also say that an actual *world* always precedes cognitive activity as its universal ground, and this means first of all a ground of universal passive belief in being which is presupposed by every particular cognitive operation .... It is this *universal ground of belief in a world* which all praxis presupposes. (EJ, 30)[17]

We thereby see how phenomenology's "obsessive" preoccupation with the interiority of consciousness culminates in the world problematic. The world as ground

---

[13] See in particular A VII 5 and A VII 10.

[14] "Das Erfassen ist ein Herausfassen." (Hua III/1, 71)

[15] See in this regard the foregoing analysis of this claim in Sect. 5.2.

[16] "Dem Erfassen liegt immer voran die Affektion, die nicht ein Affizieren eines isolierten einzelnen Gegenstandes ist." (EU, 24)

[17] "Affizieren heißt Sichherausheben aus der Umgebung, die immer mit da ist [...]. Die Umgebung ist mit da als ein *Bereich der Vorgegebenheit*, einer *passiven* Vorgegebenheit, das heißt einer solchen, die ohne jedes Zutun ... immer bereits da ist. Wir können auch sagen: aller Erkenntnisbetätigung voran liegt als universaler Boden eine jeweilige *Welt* [...] ein Boden universalen passiven Seinsglaubens, den jede einzelne Erkenntnishandlung schon voraussetzt [...]. Dieser *universaler Boden des Weltglaubens* ist es, den jede Praxis voraussetzt." (EU, 24–25)

of experience turns out to be nothing other than the ground of the *universal passive belief in being*. As such, it is nothing other than *Weltbewußtsein* itself. Thus on the one hand, it is certainly true that the world-horizon as ground of experience proves to be a determination of consciousness itself. Yet on the other hand, the world-horizon is a determination of consciousness only insofar as consciousness consists in the reaching out for that which it itself is not, in the encounter of that which transcends it. The world-horizon as ground of experience thereby bespeaks the general *style* in which consciousness discloses the world and things as intuitive.

Husserl characterizes the fundamental feature of such a universal passive belief in being as the world's self-evident (*selbstverständlich*) pregivenness. How is one to understand the self-evidence in question? To characterize the universal passive belief in being as the ground of experience implies that the belief in question is not only the ground of all judgments, but also of all modalities of experience. Being the ground of each and every modalization, it itself could not have arisen out of modalization. But if so, then it could not be given as doubtful, probable, or possible; being immune from doubt, probability, and possibility, it can only be self-evident.

Thus the world reveals itself as the ground for all modalities of experience, which Husserl addresses in §21 of *Experience and Judgment*. The world evades modalization for the simple reason that modalization is possible only insofar as it already rests upon the prior givenness of the world. This should come as no surprise: the distinction between *Weltbewußtsein* and *Welterfahrung* already implies such a scenario. The emergence of modalities is such that it presupposes the pregivenness of a simple belief, without which they could not be modalities in the first place. Something must be first given if it then is to be doubted, negated, or disclosed as probable or possible. Something must first arise, and it arises from a more rudimentary background of sense, which as Husserl remarks, each and every emerging prominence cannot help but must continue to entail. The most originary figure of the not yet modified belief, which does not only pertain to particular objectivities but also embraces the whole "region of regions" from which concrete objectivities emerge, is what I would like to call the world-horizon as the ground of experience.

Consider how objectivities emerge within, for instance, the field of vision. Their emergence is such that it point back to their previous "having been": "if I grasp in its particularity some object or other, for example in looking at a book resting on the table, then I grasp something that for me is an existent, something which, as already existing in advance, was already 'there,' 'in my study,' even though my attention was not yet directed toward it. In exactly the same way, this entire study ... with the sense 'room in the house,' the latter being on the familiar street, the street in my town, and so on" (EJ, 30–31).[18] Perceptual experience is always such that it presents objects as worldly, as emerging within an always already familiar world.

[18]"Erfasse ich in Sonderheit in meinem Wahrnehmungsfeld, z.B. auf ein Buch auf dem Tisch hinsehend, irgendein Objekt, so erfasse ich ein für mich Seiendes, das schon vorher für mich seiend, schon 'dort' war, 'in meinem Studierzimmer', auch wenn ich noch nicht darauf gerichtet war; ebenso wie dieses ganze Studierzimmer... mit dem Sinne 'Zimmer meiner Wohnung' in der vertrauten Straße, Straße in meinem Wohnort, usw." (EU, 25)

Existence, therefore, can have no other sense than that of *Inexistenz*, that of being within an already pregiven world.[19] While the pregivenness of the world as ground is to be understood as the pregivenness of sense within which each particular objectivity emerges, the most rudimentary dimensions of what is entailed in this pregivenness should be seen as an accomplishment of *Weltbewußtsein*, and not of *Welterfahrung*.

Thus Husserl's genetic analysis of transcendental logic discloses the world as the fundamental ground of experience. However, Husserl's richest account of this dimension of the world-horizon is to be found in a different context. Let us turn to an unfinished manuscript from 1934, in English known as "Foundational Investigations of the Phenomenological Origin of the Spatiality of Nature: The Originary Ark, the Earth, Does Not Move"[20] as an exemplary account of the first determination of the world-horizon.

10.4 The Earth Does Not Move

At the beginning of this unfinished piece, which according to Husserl's own estimation is "foundational for a phenomenological doctrine of the origin of spatiality, corporeality, Nature" ("FI," 117), Husserl introduces a distinction between two notions of the world—a *guiding* distinction for Husserl's further analysis. First, Husserl speaks of the world as an intersubjectively formed surrounding world which, due to the apperceptive transference of sense, stretches into infinity. The constitution of this world is rendered possible on the basis of the background/foreground schema: what is given to me directly is a particular *segment* of the world, which is from the start apperceived as a layer of an infinite whole, whose other segments either are, or can be directly and intuitively given to me or others. Secondly, Husserl also emphasizes that this first notion of the world gives rise to a number of different representations of the world—for instance, the world of the Greeks or the modern Copernican world of the natural sciences.

---

[19] Such a notion of the world as the ground of experience is of vital significance for Husserl's conception of transcendental logic. The ground of the world presents itself as a unity, which stands for the pregiven totality of what is experienceable, a totality, which in its own turn, is a correlate of the unity of experience. This is of paramount significance for Husserl: it allows him to qualify traditional logic as *the logic of the world*. No matter how formalized the "S" and the "p" of formal logic might be, the first of these terms nonetheless must stand for a formalized being, which must belong either to the actual or an imaginable world. Such must be the case because the "something" that appears shares certain elementary features with concretely given objects of experience: it is an identity in the unity of experience, and correlatively, an identity in the world. See in this regard de Almeida (1972), pp. 180–203.

[20] The original title of this manuscript is "Umsturz der kopernikanischen Lehre in der gewöhnlichen weltanschaulichen Interpretation. Die Ur-Arche Erde bewegt sich nicht. Grundlegende Untersuchungen zum phänomenologischen Ursprung der Körperlichkeit, der Räumlichkeit der Natur im ersten naturwissenschaftlichen Sinne. Alles notwendige Anfangsuntersuchungen." (D, 17)

The general goal of this manuscript is that of disclosing how the second notion of the world, exemplified by the Copernican worldview, is grounded in the first one. Yet as this analysis is carried through, it becomes also clear that the established distinction between these two notions of the world is not exhaustive in that it leaves out of account a more rudimentary pregivenness of the world. Husserl's analysis is geared toward the apprehension that even though the intuitive world is a necessary basis upon which the scientific world can emerge, it nonetheless itself already rests upon the world-horizon as the ground of experience. There is thus a *third notion* of the world operative in Husserl's analysis of the distinction between the intuitive world and a particular worldview. Husserl thematizes this third notion under the heading of the "originary ark" (*Ur-Arche*)—a notion, which bespeaks the most rudimentary givenness of the *earth*, when the latter is conceived as a ground of both particular worldviews and particular intuitive worlds.

Husserl, however, does not establish the earth as ground of experience as an essentially different notion than that of the intuitive world. He rather identifies the "originary ark" as a necessary dimension of the intuitive world itself. Yet such identification overshadows the difference between the constitutive frameworks that underlie the conceptions of the world in question. I have already addressed this issue in Sect. 10.1: while the intuitive world presupposes the foreground/background structure, the world as ground is only conceivable as a background without any foreground. Thus while the intersubjectively formed surrounding world is intuitive, thematic, and up to a degree objective, the world as ground is neither intuitive, nor thematic, nor objective. Let us identify the "originary ark" as the third notion of the world so as not to lose sight of these constitutive differences.

Let us turn back to the explicitly drawn distinction between the first and the second notions of the world. Husserl proceeds in a regressive fashion, starting with the acknowledgment of the fact that "for us, moderns" the earth is one spherical body (Körper) among others. And as he later observes, for us, after Copernicus, it is simply ridiculous to want to believe that the earth is the center of the world "simply because by accident we live on it." Yet the Copernican conception of the world clearly comes into conflict with the experiential givenness of the world: as far as we can abstract from the influence of modern science, we are obliged to assert that for us, the earth is the experiential ground of all bodies, which underlies the experiential genesis of any particular representation of the world. "This 'ground' is not experienced at first as body but becomes a corporeal ground at higher levels of constitution of the world by virtue of experience, and that nullifies its primordial form of the ground" ("FI", 118).[21] One must not lose sight of the genetic framework within which the conflict between the scientific and the intuitive worlds emerges: there can be no question that the scientific view of the world springs from the more primordial intuitive givenness of the world. And thus the question that such a genetic framework

[21] "Dieser 'Boden' wird zunächst nicht als Körper erfahren, in höherer Stufe der Konstitution der Welt aus Erfahrung wird er zum Boden-Körper, und das hebt seine ursprüngliche Boden-Form auf." ("Umsturz," 308)

gives rise to is the following one: what does the prehistory of the world's intuitive givenness mean for the scientific worldview?

Husserl's inquiry into this question takes a twofold form. First, Husserl wants to show that such a prehistory of the scientific worldview is possible only because the intuitive world lends itself to a Copernican interpretation. Secondly, Husserl also wants to argue that the success of the scientific worldview should by no means be identified with the annulment of what is entailed in the notion of the intuitive world. The Copernican worldview does not cancel the sense of the intuitive world; rather, the intuitive world continues to persist as a necessary dimension of the Copernican worldview. As we will see below, such an irreducible presence of the intuitive world, conceived as a dimension of the Copernican one, significantly curbs the pretenses of the scientific worldview. Yet first, let us take a closer look at those dimensions of the intuitive world that underlie the emergence of the Copernican worldview.

To conceive of the earth as a body, as one planet in the infinite space of the world, is to attribute to it movement, which it would then share with each and every other objectivity. Yet such an attribution comes into a direct conflict with the *how* of the world's givenness: "in the primordial shape of its representation, the earth itself does not move and does not rest; only in relation to it are movement and rest given as having their sense of movement and rest" ("FI", 118).[22] It is this notion of the earth which, to paraphrase Merleau-Ponty,[23] lies *on the other side* of rest and movement, that Husserl characterizes as the "originary ark." And it is this notion, I would like to contend, that is the ground not only of particular worldviews, but also of the intuitive world.

On Husserl's view, as long as the earth does not move, rest and movement are given as *absolute*. Yet let us ask, what kind of notion of the world is presupposed by the *absoluteness* of rest and motion? Husserl's answer points in the direction of the intuitive world. Yet the malleable character of the intuitive world is too inclusive and thus too imprecise to account for the absoluteness in question. As Husserl himself acknowledges, a certain relativity of rest and motion is to be found already within the intuitive world ("Umsturz," 311–12). I would therefore like to contend that the notion of the world that underlies rest and motion as absolute is not that of the intuitive world, but rather that of the world conceived as the non-intuitive ground of experience. *Only as long as the earth remains non-thematic, non-objective, and non-intuitive can rest and motion, as they pertain to concrete objectivities, be given as absolute.* And correlatively, the earth's becoming thematic, objective, and intuitive indicates the relativization of rest and motion.

Due to such a distinction between the world as ground and the intuitive world—a distinction which, although non-thematic, is still operative in this text—Husserl can assert that the confirmation of the scientific representation of the world finds not only its support, but even its core in my perceptual field, i.e., in my intuitive world. "Already a certain relativity of rest and motion is formed here" ("FI", 121).[24]

---

[22]"Erde selbst in der ursprünglichen Vorstellungsgestalt bewegt sich nicht und ruht nicht, in Bezug auf sie haben Ruhe und Bewegung erst Sinn." ("Umsturz," 309)

[23]See Merleau-Ponty (2002), pp. 67–76.

[24]"Eine gewisse Relativität von Ruhe und Bewegung ist hier schon ausgebildet." ("Umsturz," 311)

To employ one of Husserl's examples, I can be in a railway-car and conceive of it as my ground-body. When it carries me in motion, I experience it as though it were at rest. And thus the countryside appears to me in constant movement, even though I am fully aware that it is at rest. I am aware of such reversals of rest and motion within the intuitive world, for I do not forget that only for the time being did I leave the permanent ground for its temporary substitute. Therefore, "all of this is nonetheless directly referred to the ground of all relative ground-bodies, to the earth-ground" (FI, 121).[25]

This is a revealing example. It indicates that *I can experience the world as body only because previously I had already experienced it as ground.* More particularly, I can experience the world as body only if I transfer the sense of the earth-ground to other bodies and thus experience them as ground, from which the earth itself can then appear as a body. Yet if such is the case, then a more basic question forces itself upon us: by what means am I capable of transferring the sense of ground to other bodies, if in its originary figure there is only one ground—the earth-ground?

One can extract an answer to this question from my experience of Others. At the center of Husserl's notion of intersubjectivity lies the realization that the Other is not only *Körper* but also *Leib.* In virtue of my corporeality, I experience Others and enter into a relation with their experience. It is important not to overlook that the distinction between *Leib* and *Körper* does not exclusively pertain to other human beings. It also enables me to enter into an understanding with members of other species. Thus, to employ another of Husserl's examples, I understand a flying bird by broadening my kinesthesias. Or as Merleau-Ponty has put it in his interpretation of the "Umsturz" text: "I understand the bird as a (gliding flight) variant of my *Ich gehe.* Thereby as an extension of my experience" (Merleau-Ponty 2002, 72). Due to such a transference of sense, the earth as ground becomes relativized: it becomes a body seen from another ground.

Yet as I have already noted, my ability to substitute a new ground for the permanent earth-ground carries with it the memory of the first ground, which is not obtained in virtue of an act of substitution. However, given the limits that pertain to memory, it is not only possible, but also quite likely that the recollection of the originary ground would lie beyond our reach. Even more: I can be born a sailor's child and conceive of the ship as my own "earth," my primordial homeland ("Umsturz," 318). Or even better: the future humanity can identify a particular star or airship as its permanent homeland ("Umsturz," 319). And yet, no matter how far-reaching the substitution of the acquired homelands might be, it remains the case that all relative histories are episodes of a single primordial history. Such is the case because one can always ask the dwellers of the ship or the inhabitants of the stars: "how have you gotten there?"—"just as similarly in the case of a new island where cuneiform writing is found, I ask: how did the people in question come here?" ("FI," 130).[26]

---

[25]"Es ist doch alles zunächst auf den Boden aller relativen Bodenkörper, auf den Erdboden bezogen" ("Umsturz," 312).

[26]"So ähnlich wie ich bei einer neuen Insel, auf die ich Keilinschriften vorfinde, frage: wie sind die betreffenden Völker dahin gekommen?" ("Umsturz," 324)

Such a state of affairs not only enables us to understand how the intuitive world forms an irreducible dimension of the scientific worldview, it also allows us to see why the presence of this dimension limits the sanctimony of the Copernican worldview. The intuitive world remains irreducible because without it, the question of the emergence of the Copernican worldview would continue to hang in the air—it would remain unclear how motion and rest can be meaningful, or how they can become relativized, or, moreover, what it means to qualify the world as body, or how other bodies can become temporary homelands. Thus the intuitive world is an irreducible dimension of the Copernican worldview in that it is the *prehistory* of the formation of this particular representation of the world. Just as the recognition of my body as *Körper* does not make it lose its unique sense as primordial *Leib* from which all other bodies derive their sense, so the earth as the primordial homeland does not lose its unique sense in virtue of being recognized as a body among other bodies.

The admission that the prehistory of the scientific worldview belongs to it as its irreducible component of sense forces one to recognize certain limits that pertain to the central claims that stem from this worldview. As Husserl puts it, "one should not perpetrate the absurdity, indeed the absurdity, of presupposing tacitly beforehand the naturalistic or prevailing conception of the world—in order to consider then, in an anthropological and psychological way, the formation of science and of the interpretation of the world in human history, in the history of the species, within the evolution of the individual and people, as an obviously accidental event on the earth which might just as well have occurred on Venus or Mars" ("FI," 129–130).[27] The Copernican truth remains a *bound* truth, i.e., bound to its emergence, to its genesis. Or to put the matter in slightly different terms, it remains bound to the *horizon* of sense, without which it could have never been able to arise. As Husserl formulates it,

> There is only one humanity and one earth—all the fragments which are or have been separated from it belong to it. But if this is the case, need we say with Galileo: *par si muove?* And not on the contrary: it does not move? It is certainly not so that it moves in space, although it could move, but rather, as we tried to show above, the earth is the ark which makes possible in the first place the sense of all motion and all rest as mode of one motion. But its rest is not a mode of motion. ("FI," 130)[28]

[27]"Man darf nicht die Verkehrtheit, in der Tat Verkehrtheit, begehen, im voraus unbemerkt die naturalistische, die herrschende Weltauffassung vorauszusetzen und dann anthropologistisch und psychologistisch in der Menschengeschichte, die Speziesgeschichte, innerhalb der individuellen und Völkerentwicklung die Ausbildung der Wissenschaft und der Weltinterpretation anzusehen als ein selbstverständlich zufälliges Geschehen auf der Erde, das ebensogut auf Venus oder Mars stattbaben könnte." ("Umsturz," 323)

[28]"Es gibt nur eine Menschheit und eine Erde—ihr gehören alle Bruchstücke an, die sich ablösen oder je abgelöst haben. Aber wenn dem so ist, dürfen wir mit Galilei sagen, dass *par si muove*? Und nicht im Gegenteil, sie bewegt sich nicht? Freilich nicht so, dass sie im Raum ruht, obschon sie sich bewegen könnte, sondern wie wir es oben darzustellen versuchten: sie ist die Arche, die erst den Sinn aller Bewegung ermöglicht und aller Ruhe als Modus einer Bewegung. Ihr Ruhen aber ist kein Modus einer Bewegung." ("Umsturz," 324)

Thus *the earth as primordial homeland is neither intuitive, nor thematic, nor objective; nonetheless, it is given—and even given within the scientific representation of the world.* The earth as the originary ark remains an indispensable dimension of the Copernican worldview.

This detour to the notion of primordial earth in Husserl's phenomenology clarifies the sense of the world as the non-intuitive, non-thematic, and non-objective ground of experience. Such a notion, despite its threefold negative qualification, remains a valid phenomenological notion because its negative characterizations do not cancel its *givenness*. Now we are in the position to see how such givenness is possible. The world as the wherefrom of experience *is given as an inseparable dimension of sense that pertains to the constitutive history of the formation of each and every objectivity*. It is given as the ultimate origin from which everything intuitive, thematic and objective springs. But if so, then alongside Husserl's call to stop perpetrating the absurdity of presupposing the naturalistic view of the world while inquiring into its own formation, we need to be also alerted to another kind of "transcendental naiveté," viz., *the naiveté of tacitly presupposing the intuitive, thematic, and objective manifestation of the world before inquiring into its own crystallization.* The apparent absurdity of a world given in such a threefold negative characterization is no greater than that of the earth which does not move. Yet what they both share is not only the semblance of improbability, but also their non-derivative resourcefulness of sense to which the phenomenological analysis of sense-formation leads and from which it takes its departure.

10.5 Horizont, Hof, Hintergrund (II)

Yet if the world as the *wherefrom* of experience is a figure of horizon-intentionality, then what sense are we to make of the claims made in Part I of this study regarding the relation between the notions of *Horizont, Hof,* and *Hintergrund*? In Sect. 3.3 I argued that even though all three notions bespeak the non-intuitive co-presence that accompanies the givenness of an intuitive core, the notion of the horizon stands out from those of background and halo because it does not only accompany, but also *co-determines* the sense of what is intuitively given. Having drawn such a distinction, we were further able to see how Husserl's notion of the horizon differs from James' notion of the fringes of consciousness. James' notion is equivocal, I argued, precisely because it does not distinguish between *co-presence* and *co-determination*. Yet my present analysis calls such a distinction into question. In direct contrast to Husserl's insistence in *Ideas I*, that for a consciousness transposed into the mathematical "world," the natural world no longer functions as a horizon (Hua III/1, 60), my present analysis of the world-horizon as the *wherefrom* of experience leads to the realization that the world continues to function as a horizon even when it is non-objective, non-thematic, and non-intuitive. Thus in direct contrast to *Ideas I*, both *Hintergrund* and *Hof* once again prove to be figures of horizon-intentionality, as can be easily witnessed from Husserl's interchangeable employment of the three terms not only before, but also after the publication of *Ideas I*.

Yet such a reversal of the position in regard to the relation between *Horizont, Hof* and *Hintergrund* should not be misconceived as a return to a Jamesian view, which Husserl held prior to *Ideas I*. Far from being a return to a standpoint that is indifferent to the distinction between co-presence and co-determination, this reversal rather bespeaks a genetic deepening of horizon-intentionality. Such a deepening consists in the realization that co-determination need not rely only on the resources of co-presence, that even what is intuitively absent can nonetheless play the role of horizon-consciousness, i.e., can co-determine the sense of an appearing objectivity.

Thus from a static perspective, the world as the *wherefrom* of experience appears as the darkest *Hintergrund* or *Hof*, to which the sense of appearing objectivity is completely indifferent. A genetic deepening of such a point of view brings to light the *horizonal* sense of the world's pregivenness: despite its non-objective, non-thematic, and non-intuitive givenness, the world nonetheless continues to determine objectivities as *worldly*. Such a genetic deepening of horizon-intentionality bespeaks a peculiarity that pertains to the world-horizon and it is this peculiarity that calls for an abandonment of Husserl's claim in *Ideas I*: "the natural world is a background for my act-consciousness, but it is *not a horizon*" (*Ideas I*, 55).[29] In direct contrast to such a point of view, we are now led to the realization of an emblematic characteristic that pertains to the givenness of the natural world. When we are absorbed within the arithmetical universe, the natural world still functions *as the horizon*. The opposite, however, is not the case: as soon as the subject withdraws from the mathematical considerations, distracted, let us say, by a sudden transformation that takes place within the perceptual world, the mathematical universe quickly vanishes from consciousness. As Gurwitsch succinctly puts it, "*the perceptual world has thus the privilege of omnipresence, a privilege it shares with the stream of consciousness and the embodied existence, but with no other realm*. Except these three, there is no other order, scientific, artistic, etc., with whose existence we are at every moment in some form confronted" (Gurwitsch 1985, 40–41).

References

Claesges, Ulrich. 1972. Zweideutigkeit in Husserls Lebenswelt-Begriff. In *Perspektiven transzendentaler Phänomenologie*, ed. U. Claesges and K. Held. The Hague: Martinus Nijhoff.

de Almeida Guido, Antonio. 1972. *Sinn und Inhalt in der genetischen Phänomenologie E. Husserls*. Dordrecht: Springer.

Fink, Eugen. 1990. *Welt und Endlichkeit*. Würzburg: Königshausen und Neumann.

Gurwitsch, Aaron. 1985. In *Marginal consciousness*, ed. L. Embree. Athens: Ohio University Press.

Husserl, Edmund. 1940. Umsturz der kopernikanischen Lehre. In *Philosophical essays in memory of Edmund Husserl*, ed. Farber Marvin. Harvard: Cambridge University Press.

[29] "Bewegt sich mein cogito nur in den Welten dieser neuen Einstellungen, so bleibt die natürliche Welt außer Betracht, sie ist für mein Aktbewußtsein Hintergrund, aber sie ist kein *Horizont*" (Hua III/1, 60).

Husserl, Edmund. 1970. *The crisis of European sciences and transcendental phenomenology. An introduction to phenomenological philosophy*. Trans. David Carr. Evanston: Northwestern University Press.

Husserl, Edmund. 1973. *Experience and judgment: Investigations in a genealogy of logic*. Trans. James Churchill and Karl Ameriks. Evanston: Northwestern University Press.

Husserl, Edmund. 1976a. In *Die Krisis der europäischen Wissenschaften und die transzendentale Phänomenologie. Eine Einleitung in die phänomenologische Philosophie*, ed. Biemel Walter. The Hague: Martinus Nijhoff.

Husserl, Edmund. 1976b. In *Ideen zu einer reinen Phänomenologie und phänomenologischen Philosophie. Erstes Buch: Allgemeine Einführung in die reine Phänomenologie*, ed. Schuhmann Karl. The Hague: Martinus Nijhoff.

Husserl, Edmund. 1983. *Ideas pertaining to a pure phenomenology and to a phenomenological philosophy. First book: General introduction to a pure phenomenology*. Trans. F. Kersten. The Hague: Martinus Nijhoff.

Husserl, Edmund. 1999. In *Erfahrung und Urteil. Untersuchungen zur Genealogie der Logik*, ed. Landgrebe Ludwig. Hamburg: Felix Meiner Verlag.

Husserl, Edmund. 2002. Foundational investigations of the phenomenological origins of the spatiality of nature: The originary ark, the earth, does not move. In *Husserl at the limits of phenomenology*, ed. Merleau-Ponty Maurice, L. Lawlor, and B. Bergo. Evanston: Northwestern University Press.

Landgrebe, Ludwig. 1952. Philosophie der Gegenwart. Bonn: Athenaeum Verlag.

Landgrebe, Ludwig. 1963. Der Weg der Phänomenologie. Gütersloh: G. Mohn.

Merleau-Ponty, Maurice. 2002. *Husserl at the limits of phenomenology*. Trans. Leonard Lawlor and Bettina Bergo. Evanston: Northwestern University Press.

Sommer, Manfred. 1998. Husserl on 'ground' and 'underground'. In *Phenomenology of interculturality and life-world*, ed. R. Zhang and E.W. Orth. München: Verlag Karl Alber.

Steinbock, Anthony. 1995. *Home and beyond: Generative phenomenology after Husserl*. Evanston: Northwestern University Press.

Chapter 11
The World-Horizon as the *Wherein* of Experience

11.1 The World as Horizon of All Horizons

Let us turn back to the distinction between *Weltbewußtsein* and *Welterfahrung*, introduced in Sect. 10.1. In my foregoing analysis, I attended only to the first of these notions. My analysis thereby procured a conception of the world in virtue of a breach with the constitutive schema operative in Husserl's analysis of the horizons of objectivities. Thus I claimed that while concrete objectivities have their own inner- and outer-horizons, whose constitution can be accounted for through the background/foreground schema, the world of *Weltbewußtsein* is the givenness of a background without any conceivable foreground: it is a horizon without any intuitive core. Yet such a conception of the world is neither the only one, nor even privileged in Husserl's writings. The fundamental difference between the world of *Welterfahrung* and that of *Weltbewußtsein* rests upon a difference within the constitutive schema itself: in contrast to *Weltbewußtsein*, *Welterfahrung* presupposes the same constitutive framework of background/foreground that pertains to concrete objectivities. And it is this schema, as we will soon see, that calls for the qualification of the world as the *horizon of all horizons*.

Yet does the presence of an identical constitutive schema in the case of the world and of things not require one to concede that, despite Husserl arguing to the contrary, the world of *Welterfahrung* is, no matter how unique, still one object among others? So as to see why this question should be answered in the negative, let us turn to another manuscript that belongs to the A VII group, significantly titled "*To Systematically Disclose the Constitution of the pregiven World Is Nothing Other Than To Systematically Disclose the Structure of the World-Horizon*" (Hua XXXIX, 125).[1]

[1] "*Die Konstitution der vorgegebenen Welt systematisch auslegen—das ist systematisch die Horizontstruktur derselben auslegen.*" (Hua XXXIX, 125)

S. Geniusas, *The Origins of the Horizon in Husserl's Phenomenology*, Contributions to Phenomenology 67, DOI 10.1007/978-94-007-4644-2_11,

This manuscript represents Husserl's strategy to address the world-horizon, conceived in terms of *Welterfahrung*, on the basis of the background/foreground schema. Within such a framework, the world-horizon is not established, as it was in the domain of *Weltbewußtsein*, in opposition to the inner- and outer-horizons that pertain to concrete objectivities; rather, its establishment is directly *founded* upon the horizons of concrete objectivities. "Starting with any experienced reality and its particular horizonality, one can disclose the pregiven world .... This is the ground of the validity of being" (Hua XXXIX, 127).[2] Let us accept this phrase as Husserl's straightforward acknowledgment that the fecundity of the inner- and outer-horizons is not exhausted by the actualization of objectivity's new modes of givenness. The inner- and outer-horizons also generate the horizonality of the world. Yet how?

At the beginning of this manuscript, we encounter what at first glance seems to be merely a repetition of those elements of horizon-consciousness Husserl had already disclosed in *Ideas I*. The appearance of each and every objectivity is said to have its own horizons of sense. Such a state of affairs is further interpreted to mean that consciousness prefigures potential modes of appearance along with the actual ones. Husserl further observes, again repeating his earlier analysis, that only due to such a prefiguring can consciousness intend objectivities and not only their profiles. Yet on the basis of this general determination of horizon-consciousness, Husserl introduces a new dimension of sense that remained absent from *Ideas I*. He insists that what in virtue of the inner- and outer-horizons comes to fulfillment is not merely a new mode of appearance, *but also new horizons that embrace potential appearances*.[3] Of central significance is the realization that *horizon-consciousness, besides prefiguring a new mode of appearance of objectivities, also prefigures a new horizonality of sense*. Moreover, this prefigured horizon of sense itself entails references not only to other potential appearances, but also to *other potential horizons*. We thus face a limitless implication of potential horizons within the actual one. The implication of potential horizons is given without clearly defined limits, and thus Husserl speaks of the "endlessness of implications as an intentional structure of the total-horizon as a horizon."[4]

These endless implications of new horizons indicate the manner in which the world of *Welterfahrung* can be said to be always "apriori" anticipated. "The actual world in its possibilities is always anticipated, and anticipated in such a way, that it can be rendered intuitive in subsequent unanimity as a possibility in the full scope of its possibilities" (Hua XXXIX, 127).[5] I will turn to the problematic of possibility

[2]"Von jedem erfahrenen Realen aus und seiner eigenen Horizonthaftigkeit (in seinen Möglichkeiten) ist die vorgegebene Welt zu konstruieren[...]. Das ist der Boden der Seinsgeltung." (Hua XXXIX, 127)

[3]We find the same claim expressed in the C-Manuscripts: "die Bewährung betrifft nicht bloß das eigentlich Erscheinende und Stimmende, sondern es bewährt sich das Dasein der Welt selbst, es bewährt sich m.a.W. mit die Antizipation der Horizontmeinung." (Mat VIII, 428)

[4]"Unendlichkeit der Implikationen, also eine intentionale Struktur des Gesamthorizontes als Horizontes." (Hua XXXIX, 128)

[5]"Die wirkliche Welt in ihren Möglichkeiten ist immerfort 'a priori' antizipiert, und antizipiert derart, dass sie als eine Möglichkeit im Spielraum ihrer Möglichkeiten in konsequenter Einstimmigkeit anschaulich gemacht werden kann." (Hua XXXIX, 127)

in the next section. For now it is important to note that the implication of new horizons within the actual one is not such that they would give rise to endless contradictions and thereby cancel the very notion of an endless horizon. Husserl rather insists repeatedly that such an endless implication of horizons is given within unanimity of sense: "the horizon thus expresses unanimity" (Hua XXXIX, 128).[6] To describe how such unanimity is possible is precisely to disclose what is entailed in the notion of the world of *Welterfahrung*.

Yet before turning to a concrete elaboration of this task, let me emphasize that such a concordant implication of endless horizons within the actual one is what allows one to qualify the world as the horizon of all horizons. Borrowing this phrase from Stephan Strasser[7] (although employing it differently), I want to emphasize that the world of *Welterfahrung* precisely is what allows for the intertwining of all horizons. Appropriating Gadamer's characterization of understanding and modifying it slightly, one could say that the world is what underlies the merging of all horizons, supposedly existing by themselves.[8]

We thereby see that what the background/foreground schema gives rise to is not only the appearance of concrete objectivities. It just as much generates the implication of potential horizons within the actual one, which, once taken in its concrete endlessness, is nothing other than *Welterfahrung*. Thus in contrast to the world of *Weltbewußtsein*, the world of *Welterfahrung* is founded upon the givenness of concrete objectivities, i.e., it is founded upon the inner- and outer-horizons that bring things to their self-givenness. Yet horizon-intentionality does not bring to self-givenness only concrete objectivities. To speak of the prefiguring of potential horizons of sense is to claim that the horizons are also brought to fulfillment, i.e. brought to self-givenness. In the most direct fashion Husserl asserts that such is the case: "the horizon as a horizon comes to 'self-givenness'" (Hua XXXIX, 141).[9]

We thereby obtain the notion of the world as the horizon of all horizons. Yet this phrase still remains equivocal, and thus the present analysis can follow one of two directions. We can aim to extract what the implication of all horizons within the actual one means by disclosing the formal structure of the world as horizon of all horizons. By pursuing this direction, we can obtain a notion of the world as it manifests itself in virtue of *open* possibility. I will call this dimension of world-experience "the world as the *wherein* of experience." Yet we can also thematize the interweaving of horizons by paying heed to the concrete and constant transformations of sense that such an *Ineinander* brings forth. By following this direction, we will be led to disclose the world-horizon by taking into account other modalities of experience. I will term this dimension of world-experience "the world as the *whereto* of experience." Let me begin with the first possibility.

---

[6] "Horizont drückt also Einstimmigkeit aus." (Hua XXXIX, 128)

[7] See Strasser (1976), pp. 151–179 as well as Strasser (1969), pp. 23–46.

[8] See Gadamer (1990, 311).

[9] "Horizont als Horizont kommt zur 'Selbstgegebenheit.'" (Hua XXXIX, 141) The quotation marks here do not indicate a certain lack that pertains to self-givenness in question; they rather intimate that the self-givenness of the horizons is different from that of concrete objectivities.

11.2 The World as the *Wherein* of Experience

§29 of *Experience and Judgment*, entitled "Absolute Substrates and Absolute Determinations, and the Threefold Sense of this Distinction,"[10] is an exemplary analysis of the world as the *wherein* of experience. It should, however, be acknowledged that the title of this section is misleading: what this section establishes is not a threefold distinction between absolute substrates and absolute determinations, but rather a threefold manner in which substrates can be said to be absolute. One should not overlook Husserl's appraisal of the world as "the absolute substrate in its pre-eminent sense" (EU, 159). It is this notion of the world that I will qualify as the *wherein* of experience.

As we have seen in Sect. 8.4, Husserl pursues his analysis of prepredicative experience in terms of three levels of horizonality. Within the second of these levels, that of the inner-horizon, Husserl locates the origin of the distinction between substrates and their determinations (EU, §22). Thus the schematization of experiential givenness in terms of a substrate and its determinations cannot be said to be an accomplishment of predicative judgment; it is already a fundamental feature of prepredicative experience.

In their prepredicative as well as predicative manifestation, substrates and their determinations appear as interchangeable and thus as *relative*. What is first given as a determination can be rendered independent by way of "substratization." To employ one of Husserl's own examples, as one directs one's interest toward a flower bed as a substrate of experience, one can be affected by one of the flowers so strongly that the original interest in the flower bed recedes into the background thereby opening the space for a particular flower to become a new substrate. Similarly, what is first given as a substrate can be colligated with other objects and thus rendered a determination of a larger substrate in the flow of experience: a particular flower as an "original" substrate can subsequently reveal itself as a determination of a flower bed. Due to such transformations, anything that affects consciousness can be given as a substrate as well as its determination.

This relativity, however, does not dispense with the distinction between a substrate and its determinations as though it were an accidental feature of experience. Rather, "in any case, the explication inherent in the experience bears in itself the distinction between substrate and determination" (EJ, 133).[11] It is of central significance not to loose sight of the irreducibility of this structure of experience: in its wake, it will suffice to lay bare its *genetic* sense so as to show the limits that pertain to the relativity applicable to substrates and their determinations.

In fact, the very sense of substrates that arise out of substratification already entails that not all substrates could have originally been given as determinations. What emerges from substratification preserves its sense of having been previously given

---

[10] "Absolute Substrate und absolute Bestimmungen und der dreifache Sinn dieser Unterscheidung." (EU, §29)

[11] "Unter allen Umständen trägt das erfahrende Explizieren in sich den Unterschied von Substrat und Bestimmung." (EU, 151)

as a determination. A determination of what? Of a previous substrate. Thus previously another substrate must have been given so that its determination could subsequently undergo substratification. Yet such a regressive interchangeability of substrates and determinations could not go on *ad infinitum*. If there indeed are substrates that arise out of substratification, then we need to concede that there also are substrates that do not arise out of substratification. These latter substrates merit the name of *absolute* substrates.

Thus far Husserl's argument is only formal. So as to render it phenomenologically legitimate, one must support it with experiential evidence. Is there anything within the experiential field that could be qualified as an absolute substrate? Husserl's analysis of this question is a prime example of how phenomenology reengages traditional problems in philosophy by way of transplanting them into the experiential domain.

Husserl procures a threefold response. The first answer points in the direction of *concrete bodies*. Such a response at first glance seems implausible: after all, does not Husserl himself provide us with counter-examples that qualify concrete objects as relative substrates? I have already alluded to one illustration: a flower, to which I direct my attention in the wake of my interest in a flower bed, is a concrete body which is nonetheless given in virtue of substratification. Similarly, a particular tree can become a substrate of my attention due to the redirection of my interest from the previous substrate—a line of trees. These counter-examples, which Husserl himself provides in *Experience and Judgment*, call for a qualification of the identification of bodies with absolute substrates if this identification is to obtain phenomenological legitimacy.

It must be stressed that just as Husserl's formal discovery of absolute substrates rested upon the *genetic* sense of the distinction between substrates and their determinations, so the first qualification of absolute substrates must also be understood as a *genetic* claim. Husserl's claim should not be taken to mean that each and every body can only be experienced as an absolute substrate. It can indeed be given as a relative substrate or even as a determination of other substrates.[12] Nonetheless, it belongs to its experiential sense that it is an absolute substrate. Such an experiential sense derives from the constitutive sedimentations that belong to the givenness of each and every concrete body. Let us see why such is the case.

One can bring to light these constitutive sedimentations noetically, i.e., by glancing at what is entailed in the *directedness of consciousness* from which substrates, be they absolute or relative, are inseparable. Husserl writes: "it is, to be sure, toward relative substrates of the sphere of experience that every act of correspondent relative experience is directed, but this being-directed-toward, the start of the act of experience, is mediated by the experiential activities in which the absolute substrate in question has been explicated and where, finally, the relevant determinations have

[12]This is clearly stated in the discussion of *pluralities of bodies*, conceived as absolute substrates: "absolute Substrate, hier Körper, können als Bestimmungen fungieren, können die Funktion von Teilen, Gliedern in Ganzen, in Substrateinheiten höherer Stufe annehmen." (EU, 153) Yet one should not overlook that absolute substrates can only be given as independent determinations, i.e., to employ the language of §30 of *Experience and Judgment*, they can only be given as "pieces" and not "moments."

been substratified" (EJ, 134).[13] We thereby see that Husserl's formal answer has its direct parallel within the experiential domain. It belongs to the genetic sense of the directedness of consciousness ("Sich-darauf-richten") that an absolute substrate is the originary terminus of its *intentio*. This means that a body is an absolute substrate in the sense that *it can be an original object of the directedness of consciousness, i.e., that it is what can be immediately apprehensible, or simply and straightforwardly experienceable.*[14] The parallel between the formal argument and its experiential grounding could not be any closer.

On the basis of such a notion of an absolute substrate, could one not qualify its determinations as also absolute? Such a conclusion would be premature because in the delineated sense, pluralities of bodies can also be considered as absolute substrates (EU, 153). Yet clearly, the determinations that belong to a plurality of bodies as an absolute substrate can themselves be other absolute substrates. It is therefore necessary to distinguish between absolute substrates that can be *unities in pluralities* and those that can only be *unities as pluralities*. While the former can be determined only by means of absolute determinations, the latter can be also determined by other absolute substrates. Thus the form or the color of a particular objectivity is to be conceived as its absolute determination, while a tree that belongs to a line of trees, or a flower in a flower bed, is a relative determination.[15]

Husserl thereby attains the means to corroborate the Aristotelian insight that absolute substrates are *independent* while absolute determinations are *dependent* (EU, 156). Yet as soon as Husserl establishes this conclusion, he finds himself compelled to admit that in order to be justified, this conclusion calls for a more subtle qualification. After all, the independence of absolute substrates itself has its limits, and *it is the establishment of these limits that calls for an inquiry into the phenomenality of the world*:

> No individual body which we bring to givenness in experience is isolated and for itself. Each is a body in a unitary context which, finally and universally understood, is that of the world. Thus universal sensuous experience, conceived as proceeding in universal accord, has a unity of being, a unity of a higher order; the existent of this universal experience is the totality of nature, the universe of all material bodies. (EJ, 137)[16]

[13] "Auf relative Substrate der Erfahrungssphäre ist jeder Akt entsprechender relativer Erfahrung zwar gerichtet, aber dieses Sich-darauf-richten, der Einsatz der erfahrenden Tat ist vermittelt durch die erfahrenden Tätigkeiten, in denen das betreffende absolute Substrat expliziert und schließlich die betreffenden Bestimmungen substratisiert worden sind." (EU, 152–53)

[14] Here simplicity and immediacy are to be understood in opposition to the mediated character of experience, characteristic of colligation and substratification.

[15] Plural substrates can, however, also have their determinations as absolute: "Aber auch die mehrheitlichen Substrate haben solche Bestimmungen, die ursprünglich nur *als solche* auftreten können, und zwar abgesehen von den Bestimmungen ihrer Einzelkörper, die mittelbar auch ihre Bestimmungen sind. Es sind offenbar diejenigen Bestimmungen, welche der Mehrheit als Mehrheit eine Einheit geben, die *konfigurativen* oder *Komplexbestimmungen* im weiteste Sinne[…]." (EU, 155)

[16] "Kein einzelner Körper, den wir erfahrend uns zur Gegebenheit bringen, ist ja für sich isoliert. Jeder ist Körper in einem einheitlichen Zusammenhang, der, letztlich und universal gesprochen, der Welt ist. So hat die universale sinnliche Erfahrung, in universaler Einstimmigkeit vor sich gehend gedacht, eine Seinseinheit, eine Einheit höherer Ordnung; das Seiende dieser universalen Erfahrung ist die All-Natur, das Universum aller Körper." (EU, 156)

No concrete objectivity is isolated because "*the existence of anything real* never has any other sense than that of *existence-in*" (EJ, 34).[17] Now the qualification of individual as well as plural substrates as *existences-in* calls for a correlative qualification of the world as the *wherein* of the experiences of concrete objectivities. The world is what allows for concrete objects to stand in relation to others. While the independence of particular substrates is limited by their being-in-something (In-etwas-sein), the independence of the world is unlimited: "everything is in it, and it itself is not an in-something" (EJ, 137).[18] The being-in of particular substrates further indicates their limits, and thus their *finiteness*; correlatively, the world stands out as *limitless* and *endless*. Due to these differences, the world, as the totality of nature, merits the name of an absolute substrate in a pre-eminent sense.

Yet the world as the *wherein* of experience is a peculiar substrate. Its peculiarity hinges upon the circumstance that it does not have the same kind of determinations as concrete objectivities. A body, or a plurality of bodies, can be said to have a particular form or a particular color, yet it would be senseless to determine the world in a similar way. The reason why the world cannot have determinations of the same type rests upon its unique kind of experienciability. Rather than being of the same kind as the experience of concrete objectivities, the experienciability of the world is *founded* upon that of concrete substrates: "to be sure, the world in the sense of the totality of nature is not encountered as substrate in a simple experience; its experience is not a matter of something being simply displayed in substrate moments, in 'properties' .... But the totality of nature is also 'experienced.'" (EJ, 137).[19] One can qualify the world's "experienciability" as the bond that ties all actual and potential, individual and plural substrates. Otherwise put, the world can be qualified as the *wherein* of experience.

Thus Husserl writes: "it [the world] is the totality of existents; it is not 'in something' but is itself something total" (EJ, 137).[20] And in direct opposition to what he had stated a few paragraphs earlier, Husserl is led to assert that *only the world is independent and thus only the world is an absolute substrate in the strict sense*: everything else, including other substrates, is *unselbstständig*.

This is, indeed, a curious claim: individual and plural substrates are absolute, yet only "relatively absolute." One can, moreover, with the help of this seemingly inconsonant phrase designate not only the absoluteness of particular substrates, but also the absoluteness of the world itself. Individual and plural substrates are absolute because they are *originally* experienceable in a straightforward way (*ursprünglich schlicht erfahrbar*); they are, however, relative because they are not genuinely *independent*. They are thus absolute in regard to their determinations,

[17]"*Existenz eines Realen* hat sonach nie und nimmer einen anderen Sinn als *Inexistenz*." (EU, 29)

[18]"In ihr ist alles, sie selbst aber ist nicht ein In-etwas." (EU, 157)

[19]"Freilich ist die Welt im Sinne der All-Natur nicht substrathaft erfahren in einer schlichten Erfahrung, also nicht sich schlicht auslegend in Substratmomente, in 'Eigenschaften'[...]. Aber auch sie ist 'erfahren.'" (EU, 157)

[20]"Sie [Die Welt] ist das All-seiende, nicht 'in etwas', sondern All-etwas." (EU, 157)

yet relative in regard to the world. The world, on the other hand, is absolute in terms of its independence, yet relative in terms of its non-originary experienciability: "its absoluteness lies on its independence, but it is not an *original* substrate in the sense that it could become as a whole simply the theme of an act of simple apprehension" (EJ, 139).[21] Such a curious state of affairs makes it clear that *if by absolute substrates one is to understand the givenness of that which is both originary and independent, then there are no absolute substrates.* One can thus say that, due to the tensions between original experienciability and independence, *absolute substrates can exist only in the plural.* Thus just as in the case of the foregoing analysis of the absoluteness of consciousness in the first part of this work (Sect. 5.6), we once again need to "relativize" the absolute in Husserl's phenomenology.

Yet so far my account has not exhausted the different senses that pertain to the notion of absolute substrates in Husserl's phenomenology. Besides identifying bodies and the world conceived in terms of the totality of nature as absolute substrates, Husserl also suggests that in a looser sense, founded objectivities, such as human beings, animals, cultural objects, or useful tools, can also be called absolute substrates. They merit such a qualification because despite their being founded in more rudimentary sensuous experience, they nonetheless are originally experienced as substrates and not as determinations (EU, 158, 160).

Husserl's admission that such founded objectivities can also be called absolute substrates raises serious concerns in regard to the determination of the world as *All-Natur*. After all, Husserl was led to the inquiry into the world as the totality of nature due to the recognition of *Inexistenz* as a fundamental feature of all other substrates. Yet should such *existence-in* not also pertain to these founded objectivities, which fall into the third group of absolute substrates? That is, does this third sense of absolute substrates not call for an admission that there is a good sense in which one should speak of yet another type of an absolute substrate, viz., of the world as the *All-etwas of the founded objectivities*? Or to formulate the problem differently, does the admission of the third group of absolute substrates not necessitate one to concede that the world as *All-Natur* is determined too narrowly? Paradoxically, the world as the *wherein* of experience is presented in *Experience and Judgment* in such a way that it admits of objectivities that fall outside it. It is, one could say, a non-inclusive totality. Thus as we have already seen, Husserl qualifies the world as not "in etwas," but rather as *All-etwas*; yet would the admission of a new notion of the world, which would embrace these founded objectivities, not render the world as *All-Natur* precisely *in-etwas*, viz., as a dimension of sense within a more embracive world-experience?

Even more curiously, Husserl himself admits that the notion of the world as the totality of nature is an abstraction: "but the world of our experience, taken concretely, is not only the totality of nature" (EJ, 138).[22] Why, then, does Husserl not

[21] "Ihre Absolutheit liegt in ihrer Selbständigkeit, nicht aber ist sie *ursprüngliches* Substrat in dem Sinne, daß sie […] Thema eines schlichten Erfassens werden könnte" (EU, 159).

[22] "Aber die Welt unserer Erfahrung, konkret genommen, ist nicht nur All-Natur." (EU, 158)

procure a more robust notion of the world? Given Husserl's own admission of the reductive character of the world as *Universum aller Körper* and the need for a more embracive notion of the world that is called for by the third group of absolute substrates, it is hard to comprehend the refusal to further pursue the inquiry into absolute substrates. One can make sense of such a state of affairs by corroborating Jean Wahl's critical comment in his remarkable "Notes of the First Part of *Experience and Judgment* by Husserl": "on this point, it is nonetheless true that he [Husserl] comes to a conception of pure, universal nature, as in Galilean science, whereas certain of his other indications permit us to glimpse, or even to fully see, a totally different point of view" (Wahl 1981, 188). In direct contrast to his critical analysis of the notion of the Copernican science in "Foundational Investigations of the Phenomenological Origins of the Spatiality of Nature," within the present context Husserl generates only a transcendental justification for the scientific inquiry into nature. Such a justification might well be called-for; yet the phenomenality of the world, as the *wherein* of experience, does not justify the straightforward manner in which it is carried through.

Yet as one acknowledges it, one should not overestimate this shortcoming of Husserl's analysis. Husserl's analysis of the world as *Universum aller Körper* is phenomenologically legitimate within the circumscribed limits that pertain to absolute substrates in the manner of their givenness within sensuous experience. And even though the framework of *Experience and Judgment* does not take into account the *In-Existenz* of founded objectivities that fall into the third group of absolute substrates, this does not mean that such an account cannot be extracted from Husserl's other works. As we will see, the notion of the world, conceived in terms of the *whereto* of experience, is precisely meant to disclose the sense in which founded objectivities could also be claimed to be *worldly*.

Yet before turning to the analysis of this dimension of world-experience, a few more words need to be said in support of the claim that Husserl's notion of the world as *All-Natur* is justifiable within the limits of sensuous experience. To do so, I need to turn to a possibly far-reaching objection, according to which Husserl's notion of the world as the horizon of all horizons transgresses the boundaries of phenomenological justification in that it aims to unify what for good reasons evades unification.

11.3 World-Horizon as Open Possibility

In Jean Wahl's commentary on Husserl's *Experience and Judgment*, to which I referred above, we find a compelling formulation of a position, according to which Husserl's notion of the world as the *wherein* of experience exceeds those phenomenological boundaries that his analysis must presuppose if this notion of the world is to obtain phenomenological justification:

> Doesn't the idea of the affirmation of the unity of the world as totality of what can be validated in experience, which contains not only experiences of fact, but also possible experiences of phantasy, also seem to go beyond the frame of phenomenology? There are

> two ideas here, that of totality, and that of possibility, which must be examined. Is there a totality of everything verifiable? Is there a possible world integrated in which is such or such phantasy? Aren't these ideas which go beyond pure experience? As does, similarly the idea of the unity of our experience. (Wahl 1981, 188)

These critical notes bring into question the phenomenological legitimacy of the notion of the world as the horizon of all horizons. As we see, this notion must incorporate two dimensions of sense: *totality* and *possibility*. Yet according to Jean Wahl, such incorporation is unattainable: possibility inevitable breaks apart any kind of preconceived totality. Thus Husserl's concept of the world-horizon appears self-defeating precisely due to its ambition to be all-inclusive. As a genuine totality, it cannot retain its unity; as unified, it can no longer embrace what it claims to enfold. We thus seem to be faced with a contradiction in terms.

Yet such a critique is compelling only for as long as the precise sense of the notion of possibility remains indeterminate. That is, Jean Wahl's critique holds its sway only so long as the notion of possibility is equated with *problematic* possibility. However, Husserl's trenchant contribution to the philosophical thematization of possibility consists in heralding a distinction between *open* and *problematic* possibility (EU, §21). As we will see further on, the notion of possibility that is indispensable for the world as the horizon of all horizons is that of *open* possibility. Let us take a closer look why such is the case.

In *Experience and Judgment*, when Husserl introduces the notion of possibility into his account of the genesis of modalization within prepredicative experience, the notion he thematizes is that of *problematic* possibility (EU, §21b). Husserl refers to such a possibility as the noematic correlate of doubt-consciousness. To employ Descartes' example, to which Husserl refers starting with his earliest and ending with his latest works, we can see in a display window a given content, which we first apperceive as a human being. Yet our certainty in regard to this apperception can waver and we can be drawn to wonder if what we see is not a dressed puppet. When neither of the apperceptions takes the upper hand, our experience is that of doubt. Thus from the noetic side, consciousness of doubt is an experience of conflict between two apperceptions in which neither of the apperceptions has the power to annul its contrary. From the noematic side, what is apperceived in both conflicting cases appears as *possible*. It is here, Husserl suggests, that we are to locate the origins of the notion of possibility in the domain of prepredicative experience. Yet to this he adds a crucial remark: "in this case, these are *problematic possibilities* which are in conflict with one another" (EJ, 95).[23]

It must be admitted that if such a notion of possibility constituted an indispensable dimension of the world as the horizon of all horizons, then we would need to conclude that we face here a self-defeating notion. World-experience could then no longer be said to be an experience of unanimity, but on the contrary, an experience of irresolvable strife. It could not even be said to be the "night in which all cows are black." Rather, it would be a night which is also a day; it would be a consciousness

[23] "Es sind in diesem Fall *problematische Möglichkeiten*, die miteinander im Streit liegen." (EU, 104)

of all objects at once given in all their possible shades of colors; it would be a non-harmonious unification of the conflicts that these differences generate—a resolution of them which nonetheless leaves them unresolved. In short, evading all conceivable unanimity, the world as the horizon of all horizons could no longer be a notion at all. Such a "notion" would be completely fortuitous: its presence would add nothing to its absence.

Yet as I have already remarked, what I have thematized so far is not the only notion of possibility that we find in Husserl's work. Husserl repeatedly emphasizes the inimitable character of what he terms *open* possibility: "this possibility designates a *kind of modalization completely different from problematic possibility*" (EJ, 98); "it is absolutely essential, therefore, to distinguish the *modalities arising from conflict* and the *modalities of open particularization*" (EJ, 99).[24] One can intimate the difference between them by pointing to the different origins from which these two types of possibility spring: while the origins of problematic possibility lie within the intuition of a conflict between different apperceptions, the origins of open possibility lie within the *indeterminate generality that is given in unmodalized certainty*. While the origin of problematic possibility lies within *intuitive* consciousness, the origin of open possibility lies within *empty* consciousness. Let us take a closer look at the emptiness in question.

The empty consciousness of which I speak is nothing other than the indeterminate horizon-consciousness co-given along with any kind of intuitive core. It is a general anticipation of an indeterminate kind of fulfillment, which plays an irreducible role within perceptual consciousness. Consider seeing an object for the first time—for instance, in an exhibit of Egyptian art, a painted golden mask of Tutankhamun. Seeing the numerous colors of this mask from a certain perspective brings forth an expectation that the mask's unseen side will also reveal itself in certain colors. Yet what is entailed in the expectation is only a consciousness of "any color at all." What we face here is not an experience of conflict, as was the case in problematic possibility, but only a consciousness of indeterminate generality—a consciousness of an expectation given with an amplitude which admits of a variety of intuitions that could play the role of fulfillment.

We are, however, in need of a different example so as to see how the fulfillment of open possibility in fact is structurally identical with the fulfillment of world-consciousness addressed in Sect. 11.1. Consider turning your gaze away from the mask of Tutankhamun and noticing in front of you a colored obelisk. Seeing the colors of one of its four sides gives rise to a general expectation that a different side of the obelisk will also reveal itself in certain colors. Yet what exactly comes to fulfillment when one makes a few steps in one direction and a different side of the obelisk reveals its concrete colors? What comes to fulfillment is not only the general expectation, which culminates in the concrete intuitive givenness of particular colors;

[24]"Sie bezeichnet eine *total verchiedene Art der Modalisierung gegenüber der problematischen Möglichkeit*" (EU, 107–108). "Grundwesentlich sind demnach zu scheiden die *Modalitäten aus Widerstreit* und die *Modalitäten der offenen Besonderung*." (EU, 108)

a new dimension of sense enwraps the whole obelisk, i.e., *a new horizon of sense also comes to fulfillment*—a horizon which further anticipates the colored manifestation of the obelisk's still unseen sides. Thus just as in the case of the foregoing discussion of the world as the horizon of all horizons, we once again see that fulfillment is necessarily *double*: both intuitive and horizonal.

Yet this structural affinity of the double fulfillment in the case of concrete objectivities and of the world-experience still conceals an important difference. In the case of the obelisk, the horizonal fulfillment is to be understood as the emergence of a new expectation of sense that pertains to the manifestation of the same objectivity: the new horizon of sense that comes to self-givenness is nothing more than a horizon of sense that belongs to the obelisk. Yet what we say of the obelisk, we cannot say of the world: as already remarked, it is plainly meaningless to attribute to it shape or colors. But if qualities of concrete objectivities do not play a role within the world as the horizon of all horizons, then what could it mean to speak of the double fulfillment in regard to the world as the *wherein* of experience? In which sense can the horizon, conceived as the world-horizon, itself come to fulfillment?

In this regard, Husserl's discussion of open possibility is of decisive significance. Husserl indicates that the content of empty consciousness lends itself to its own explication in terms of its intuitive presentifications (*veranschaulichende Vergegenwärtigungen*). This means that I can run through all the possible color configurations that the golden mask of Tutankhamun or the obelisk can assume. Due to such free variation, what is entailed in empty consciousness thereby comes to *quasi-fulfillment*. The object in question thereby manifests itself to me "intuitively" in an almost infinite set of "manifestations," each of which can play the role of an actual fulfillment.

Thus intuitive presentification, accomplished through free variation, discloses the *amplitude* of empty consciousness. Yet the *Spielraum* in question need not be confined to the possible manners of manifestation that pertain to a concrete objectivity. Nothing stands in the way of a performance of such intuitive presentifications in regard to any appearing objectivity. Such a free variation does not merely bring to light the different dimensions that belong to these empty spaces of sense. It is rather the case that there is always something in common that such empty consciousnesses share. One can therefore say that *free variation also discloses the very space of sense that is entailed in consciousness conceived as open possibility.* To reveal the formal framework of what is irreducible in these repeated intuitive presentifications is nothing other than to disclose the world as the *wherein* of experience.

Now we can see how the world as the horizon of all horizons can be said to come to fulfillment. It obtains self-givenness as the domain of open possibilities taken in its broadest generality. As my interest in the obelisk recedes into the background, something else emerges into the foreground, and even though it is senseless to assume its color, sound, shape, or taste, it still remains the case that the emergent objectivity is *worldly*. From the very start, everything that affects consciousness is given in an objective apprehension: whether it is an image of an Egyptian mural, a scent of coffee, the sound of the siren of a passing ambulance, or the sight of a bored and sleepy museum guard, it is always something or someone already entailed within the world-horizon. Thus the world-horizon, conceived as the domain of open possibilities taken in their broadest generality, is a horizon of the indeterminate and the unknown, yet it already entails an element of familiarity. "The apprehension 'object in general'—still completely indeterminate and unknown—already entails an element of familiarity, namely, as a something that 'somehow or other is,' that is explicable and can be known in conformity with what it is, i.e., as something which is situated within the horizon of the world considered as the totality of existents, something which itself is already familiar *insofar as* it is a being 'in the world' and, correlatively, a being which must enter into the unity of our flowing experience" (EJ, 38).[25]

The world as the *wherein* of experience thus stands for our consciousness of the most general *types* of experience, which through their apperceptive projections of sense embrace each and every appearing objectivity. In this regard, the horizon of all horizons can be qualified as the *totality of typification* (*Totalitätstypik*) that, as Husserl himself puts it, belongs to the total horizon of the world in its infinity (EU, 33). But if so, then we need no longer be puzzled with regard to how the world as possibility relates to the world as totality. Jean Wahl's critique notwithstanding, conceived in terms of its most general indeterminacy, the word as *open possibility* reveals itself precisely as *Totalhorizont*. One can therefore qualify the world, conceived in terms of the *wherein* of experience, as the most general empty consciousness (*Leerbewußtsein*)—i.e., as consciousness of open possibility taken in its all-inclusive generality.[26]

Yet such a characterization of the world as a consciousness of the most general types of experience already entails dimensions of sense that extend beyond the limits that pertain to the world as an absolute substrate. As I have noted, this notion of the world should be conceived within the strict boundaries of unmodalized sensuous experience. The types of experience, however, are neither only unmodalized, nor only sensuous: whether I recognize the sounds I hear as a performance of an unknown piece of music or as a conversation in an unfamiliar language, whether I apperceive a particular object as a work of art or as a useful tool—these types of

---

[25]"Die Auffassung als 'Gegenstand überhaupt' – noch in völliger Unbestimmtheit, Unbekanntheit—bringt also schon ein Moment der Bekanntheit mit sich, eben als eines Etwas, das 'irgendwie ist,' das explikabel ist und nach dem, was es ist, bekannt werden kann; und zwar als ein Etwas, das im Horizont der Welt als der Allheit des Seienden darinsteht, das selbst schon so weit bekannt ist, daß es Seiendes 'in der Welt' ist[…]. (EU, 34–35)

[26]This is the sense of world-experience that underlies Husserl's proclamation in *Formal and Transzendentale Logik*: "Vor allem Urteilen liegt ein universaler Boden der Erfahrung, er ist stets als einstimmige Einheit möglicher Erfahrung vorausgesetzt. In dieser Einstimmigkeit hat alles mit allem sachlich zu tun… So hat jedes ursprüngliche Urteilen in seinem Inhalt und so jedes zusammenhängend fortschreitende Urteilen Zusammenhang durch den Zusammenhang der Sachen in der synthetischen Einheit der Erfahrung, auf deren Boden es steht." (Hua XVII, 186) We find the same claim in *Experience and Judgment*: "[…] es bleibt die nie ausdrücklich gewordene Voraussetzung, daß dieses Etwas, das eingesetzt wird, eben Seiendes ist, das eingeht in die Einheit der Erfahrung, korrelativ in die Einheit der Welt als der Allheit des Erfahbaren überhaupt." (EU, 36)

objectivities are worldly, yet they are not only sensuous. We are thus in need of a more robust notion of the world-horizon that could account for the presence of non-sensuous types of objectivities. With this in mind, let us turn to the world-horizon, conceived as the *whereto* of experience.

References

Gadamer, Hans-Georg. 1990. *Wahrheit und Methode. Grundzüge einer philosophischen Hermeneutik*. Tübingen: J.C.B. Mohr (Paul Siebeck).

Husserl, Edmund. 1973. *Experience and judgment: Investigations in a genealogy of logic*. Trans. James Churchill and Karl Ameriks. Evanston: Northwestern University Press.

Husserl, Edmund. 1999. In *Erfahrung und Urteil. Untersuchungen zur Genealogie der Logik*, ed. Landgrebe Ludwig. Hamburg: Felix Meiner Verlag.

Husserl, Edmund. 2006. In *Späte Texte über Zeitkonstitution (1929–1934). Die C-Manuskripte*, ed. Lohmar Dieter. Dordrecht: Springer.

Strasser, Stephan. 1969. *The idea of dialogical phenomenology*. Pittsburgh: Duquesne University Press.

Strasser, Stephan. 1976. Der Begriff der Welt in der phänomenologischen Philosophie. *Phänomenologische Forschung* 3: 151–179.

Wahl, Jean. 1981. Notes on the first part of experience and judgment by Husserl. In *Apriori and world. European contributions to Husserlian phenomenology*, ed. and Trans. William McKenna, Robert Harlan, and Laurence Winters. The Hague: Martinus Nijhoff.

Chapter 12
The World-Horizon as the *Whereto* of Experience

12.1 The Backward Reference of Modalized Experience

The world as the *wherefrom* and the *wherein* of experience does not exhaust the full sense of Husserl's notion of the world-horizon. The insufficiency of these notions can be grasped by way of identifying the common element that they both share. Whether one thematizes the world as ground (*Boden*) or as an absolute substrate, in both cases one aims to extract the phenomenality of the world from within *sensuous* experience taken in its *unmodalized* manifestation.[1] Modalization, however, does not take place outside the world. But if so, then phenomenology faces the task of describing and evaluating the significance of the constitutive accomplishments of modalization for world-experience. It will soon become apparent that taking modalization into view also calls for an account of the role that non-sensuous experience plays within the world-horizon. The guiding sense of the notion of the world as the *whereto* of experience thereby becomes apparent: this notion aims to describe in its essential features the enrichment of sense that the world-horizon undergoes due to *modalization* and *non-sensuous experience*.

Unmodalized sensuous experience is a progressive and unobstructed attainment of the new modes of givenness of objectivity. Such a continuous fulfillment is not a blind striving, but rather a fulfillment of protentional anticipations, i.e., of objective

[1] One could, of course, wonder if there could be anything genuinely unmodalized for Husserl. After all, in the framework of this phenomenology, experience is analyzed in terms of its givenness through neutrality *modification*. Should one not, therefore, claim that unmodalized experience lies beyond the reach of this phenomenology? Such indeed is the case, *yet only in part*. Within this context, it is helpful to recall the distinction Eugen Fink draws between the *constituting I* and the *phenomenologizing I* (Fink 1988, §6). While in the case of the phenomenologizing I, all experience is modalized from the start, the analysis of the constituting I leads to the recognition that all modalization arises from the givenness of non-modalized experience. The question of the origins of modalization does not lie outside the scope of genetic phenomenology; it is rather one of the most central questions this phenomenology investigates.

S. Geniusas, *The Origins of the Horizon in Husserl's Phenomenology*, Contributions to Phenomenology 67, DOI 10.1007/978-94-007-4644-2_12,

horizons of sense that apprehend and foreshadow the subsequent manifestation of objectivity. Thus when I see from far away a stranger waving at me, I anticipate that as his appearance becomes more perspicuous, his so-far indistinct presence will reveal a number of already familiar features. Yet it could happen so that my aspiration to gain a better glance of this presumed acquaintance is hindered by the intuition of his fixed stare, by the rattling of the mechanical movements of what now manifests itself as his artificial arms, in short, by the realization that what I thought was an acquaintance in truth is a wax doll. My unmodalized sensuous experience can always culminate in frustration; my protentional anticipation can always be obstructed by an appearance that comes into direct conflict with the horizons of anticipated sense. Within the experience of such conflicts, Husserl locates the origin of the modalities of judgment.

Let us consider the origin of negation. What happens when my seeing of an acquaintance turns out to be a seeing of a wax doll? The answer seems to be straightforward: a *disappointment* of the horizon of anticipated sense takes the place occupied by fulfillment in unobstructed perception. Husserl himself provides such an answer (EU, 94), yet he quickens to show that the distinction between disappointment and fulfillment is not as clear-cut as this characterization might suggest. It is important not to overlook that disappointment does not pertain to everything implied in the anticipated sense, for otherwise we could no longer speak of one single intentional process *within which* disappointment emerges. Thus Husserl argues that the disappointment of the intended sense itself emerges within "a unity of an intentional process," within the "unanimity of one consciousness," "within a unitary framework of sense" (EU, 95). Disappointment pertains to *certain aspects* of the intended sense and therefore leaves the unity of the anticipated sense intact. Only due to such unimpaired fulfillment can one say: *this phenomenon, this entity, or this individual I see* is not an acquaintance at all; *it* is a wax doll. The newly constituted objective sense "throws its opponent out of the saddle" (EU, 95), yet only on the basis of a more rudimentary continuous fulfillment.

Thus within a unitary horizon of anticipation, a new objective sense "wax doll" overpowers the previous one, "an acquaintance." Yet Husserl immediately adds, "but the certitude which has been overcome is still present to consciousness, although with the character of the 'null'" (EJ, 89).[2] The new objective sense, as a new phase of perception, thereby continues the preceding series of appearances. Such a continuation brings forth a curious *doubling*, which Husserl calls "a certain *doubling in the total sense-content of the perception*" (EJ, 89).[3] Just as in the streaming forth of perception, the new objective sense *invalidates*, yet does not *eradicate* the previous one, so this new objective sense also *streams back* into the retentional sphere and retroactively *cancels*, yet does not *erase* the anterior anticipation of sense. What we face here is a peculiar backward reference of negation: the antecedent apperception

[2] "Als überwältigte ist sie nun noch bewußt, aber mit dem Charakter des 'nichtig.'" (EU, 95)

[3] "Freilich erfolgt dabei eine gewisse *Verdoppelung im gesamten Sinnesgehalt der Wahrnehmung.*" (EU, 96)

"an acquaintance" in the stream of perception is retroactively reinterpreted as a "wax doll." Such a phenomenological description of the origin of negation finds its confirmation in recollection: when I now remember having seen an acquaintance waving at me, I am in possession not only of the antecedent horizon of anticipation as it was then motivated, but also of the modified realization that is built over it: what I then saw was not an acquaintance at all.

From the analysis of the modality of doubt undertaken in the last section, one can extract a similar kind of backward reference. Consider opening your eyes after deep sleep and immediately catching sight of the curtains being moved by the wind flowing in through the open window. Within this amorphous material, one can see a number of distinct shapes, be they faces, figures, even whole scenes. The content of the dream from which one has just awoken and the content of what one actually sees interpenetrate in an undecipherable way. As one is still waking up, one can wonder from one image to the next one without any certainty regarding the actuality of any of these images. Before any of these images wins the upper hand, one can have an experience of doubt. And just as in the case of negation, so in this case too each new emergence of a novel figure streams back into the former experience, transforming it, although not eradicating it. Each new experience thereby doubles the content of the unitary intentional process, and in a double sense: the former content proves dubious in relation to the new one in the streaming forth of experience; yet it also proves dubious retroactive, in the new content's streaming back into the retentional sphere.

Yet the modalities of judgment in the originary pre-logical emergence do not exhaust the amplitude of senses in which pre-predicative experience admits of a backward reference. As Husserl himself puts it, "there is an original passivity not only of sensuous givens, of 'sense data,' but also of feeling and, in contrast to this passivity, there is an active, objectivating orientation, not only in perception, but also in evaluation and in pleasure" (EJ, 71).[4] In the case of feeling-intentionality or value-intentionality, modalization also brings forth a similar kind of doubling of sense. When I realize that the other's anger toward me in an expression of concealed distress, my reaction toward the bitter words I hear can undergo a modification both in its present phase as well as in the retentional domain. Similarly, when I realize that a particular act I had considered praiseworthy was nothing more than a flavorless expression of a blind self-interest, I experience a similar overlap of senses and thus a similar backward reference as in the other considered cases.

Yet no matter how heedful Husserl's description of this curious doubling in the different forms of modalization might be, it still suffers from a significant limitation. The limitation in question is analogous to the one we were led to acknowledge in Husserl's analysis of the horizon in *Ideas I*. Just as the horizon in this work was analyzed only as it pertains to objectivities and to lived-experience, yet not as

[4]"Es gibt also nicht nur eine ursprüngliche Passivität sinnlicher Gegebenheiten, sondern auch des Fühlens, und im Gegensatz dazu nicht nur eine objektivierende aktive Zuwendung[...], sondern auch eine solche im Werten, im Gefallen." (EU, 73–74)

world-horizon, so the doubling and the backward reference of which Husserl here speaks also refer to objects and to lived-experience, yet not to the world. And just as we were led to deepen Husserl's analysis of horizon-consciousness on the basis of his subsequent investigations, so here also, the limitation in question can be overcome on the basis of the resources of Husserl's phenomenology.

In all the analyzed cases, there still remains an undisclosed sense in which one is in full right to speak of a doubling and of backward reference. So as to disclose this concealed sense, let us recall Husserl's analysis of the double fulfillment, as addressed in Sect. 11.1. As we have already seen, in this manuscript Husserl argues that each intuitive act brings forth a double fulfillment: on the one hand, each particular mode of appearance is a fulfillment of an anticipated horizonality of sense; on the other hand, a new horizon of sense itself comes to fulfillment with it. On the basis of such a double fulfillment, I was led to assert that the world is the horizon of all horizons, i.e., the implication of all horizons that each of them entails. However, this assertion was built upon unmodalized experience. We now need to inquire into the role that modalization plays within such a framework.

More precisely, we need to take into account the *productivity* of modalization. Admittedly, modalization already manifests itself as productive in terms of its doubling and its backward reference, of which Husserl speaks in *Experience and Judgment*. Yet even though such a manifestation of productivity retroactively transforms the intended sense, it hardly augments the amplitude of the horizon. It could just as well have been the case that from the very start I had interpreted this objectivity as a wax doll and not as a stranger, as a curtain and not as a particular figure or scene, etc.

What renders modalization genuinely productive is its ability to generate essentially new types of experience. When I visit a foreign country I had so little knowledge about before, I see a new habitat, new types of animals and people, I hear a new language, I come into contact with new customs. Of course, what I face is, as Husserl puts it, a new horizon of "determinate indeterminacy," and thus I carry with me a certain pre-understanding of a "habitat," an "animal," a "person," a "language" or "customs."[5] Yet this recourse to my former experience and to the pregiven conception of the world as the *wherefrom* and the *wherein* is not robust enough to account for the constitutive accomplishments that such experiences generate. Thus at any given time I can return to my native land, yet once I then encounter what I formerly had been so accustomed to, it is no longer what it used to be before. The experience of the new, in virtue of its backward reference, has already flown into the retentional domain and thereby transformed it: *it has transformed not only concrete objectivities but my world itself.*

[5]"Man is certainly a stranger to man, but always similar. When we land in a totally foreign country, as I did a few years ago in China, we feel that in spite of the greatest change of elements we have never left mankind.... When one has penetrated to the depths of singularity, one feels that it is harmonious with every other in a way that cannot be put into words. I am convinced that a progressive Islamic or Hindu world in which old ways of thinking would inspire a new history, would have with our European culture and civilization that specific affinity that all creative men share." (Ricoeur 1965)

Thus not only concrete objectivities but the world itself undergoes an enrichment of sense in virtue of backward reference. Arguably, Husserl does not thematize what such a notion of the world amounts to in his works on transcendental logic because of methodological reasons: the implied notion emerges on the other side of unobstructed sensuality and thus lies beyond the identification of the world as an absolute substrate, which must be conceived as a necessary presupposition of formal logic. And yet, the phenomenological understanding of the world-horizon remains deficient for as long as the constitutive accomplishments of modalization remain unaccounted for.

Husserl's analysis of the double fulfillment provides the necessary resources to account for these constitutive accomplishments. The new horizon of sense that comes to self-givenness through modalization does not merely overlap with the foregoing horizon of anticipation, as proved to be the case in unmodalized experience. Rather, the new horizon of sense enriches the antecedent one by either specifying it or generalizing it. Such a modification of horizonal sense is, as Husserl puts it in another manuscript, "an enrichment of the world through new intentions and acquisitions."[6] Moreover, in virtue of the backward reference, this enriched sense streams back into the retentional domain, thereby doubling the antecedent world-experience. Due to such a backward reference, I experience in full force that my native tongue is only one linguistic system among many, just as my native habitat and customs are only one specific form of life.

We thereby obtain the sense of the world as the *whereto* of experience. This notion of the world-horizon bespeaks the enrichment of sense that world-experience undergoes in virtue of new and unforeseen experiences. This notion thereby encapsulates the sedimented dimensions of sense from which my world-horizon is no longer separable. The world as the *whereto* of experience thereby accounts for how the world as the *wherefrom* and the *wherein* can concretize itself into a particular *worldview*. We now need to take a closer look at how the generation of such a world-experience is possible.

12.2 World-Experience and the Historicity of Subjectivity

One might be willing to interpret the notion of the world that arises in virtue of modalization as though it pertains only to those peculiar cases in which concrete intuitions break apart the anticipatory horizons of sense. Yet the world as the *whereto* of experience cannot be reduced to what is merely irregular, atypical, or extraordinary. On the contrary, only by means of a forceful abstraction can one obtain experiences that do not already presuppose this notion of the world-horizon. My insistence that the world as the *whereto* of experience stems from modalization should be seen as a *genetic* claim: it pertains to those sedimented dimensions of sense whose

[6] "Bereicherung der Welt durch neue Vorhaben und Erwerben." (Hua XXXIX, Text Nr. 41)

genesis leads back to modalized experience. Such a genetic framework indicates that the world as the *whereto* of experience is a *concrete* manifestation of the world whose sense in inseparable from the *historicity of subjectivity*.

If the world as the *whereto* of experience does not pertain to the peculiarities and idiosyncrasies of consciousness but rather is a notion which takes into account how the world manifests itself in its full concreteness, then this notion must in a significant way enrich the foregoing analysis of unmodalized sensuous experience. As we have seen, in such experiences, what comes to fulfillment progressively *confirms* the anticipatory horizons of sense. Thus I suggested that as the mask of Tutankhamun or the colored obelisk continuously reveal their so-far unseen features, these new manifestations confirm and thereby bring to self-givenness the anticipatory horizon of sense itself, which thereby continues to guide objectivity's dynamic self-disclosure. Yet in my foregoing analysis I did not raise the following question: wherefrom do the anticipatory horizons themselves derive their sense? In virtue of what kind of structures does it always occur that as soon as I notice a particular appearance, I immediately recognize it as an appearance of a particular objectivity? After all, not always have I been able to exclaim, as my sight would come across *this* appearance, "that's an Egyptian obelisk," to say nothing of "that's a mask of Tutankhamun!" And yet, my analysis of the world as the *wherein* of experience led to the conclusion that the world as the horizon of all horizons is a totality of typification. Now we see that we are always in possession of types of experience that cannot be accounted for by simply pointing to the pregivenness of the world. We thus see that the qualification of the world as a totality of typification tacitly presupposes those dimensions of sense that pertain to the world as the *whereto* of experience.

More particularly, the examples employed in the analysis of the world as the *wherein* of experience tacitly presupposed the *productivity* of experience. Husserl alludes to the productive constituents of experience in *Experience and Judgment*, in the context of a brief discussion of typifying consciousness: "with each new object constituted for the first time (genetically speaking) a new type of object is permanently prescribed" (EJ, 38).[7] Husserl, moreover, also insists that this genesis of new types of experience turns out to be inseparable from the world-horizon: "thus, our pregiven surrounding world is already 'pregiven' as multiformed, formed according to its regional categories and typified in conformity with a number of different special genera, kinds, etc" (EJ, 38).[8] Due to such streaming-in of experience into the world-horizon, the empty consciousness of the world can never be a consciousness of mere data. It is rather always already a consciousness of *determinate indeterminacy*, and thus what consciousness grasps as emerging, it can from the start apperceive as not only an object that lends itself to explication, but more concretely, as a particular thing, a person, a work of art, etc. The realization that each new

[7] "Mit jedem neuartigen, (genetisch gesprochen) erstmalig konstituierten Gegenstand ist ein neuer Gegenstandtypus bleibend vorgezeichnet, nach dem von vornherein andere ihm ähnliche Gegenstände aufgefaßt werden." (EU, 35)

[8] "So ist unsere vorgegebene Umwelt schon als vielfältig geformte 'vorgegeben', geformt nach ihren regionalen Kategorien, und nach vielerlei Sondergattungen, Arten usw. typisiert." (EU, 35)

experience generates a new type of experience thereby accounts for how any particular manifestation immediately awakens a horizon of determinate indeterminacy: what I see, I therefore immediate recognize as a scent of coffee, as a melody of an already familiar piece of music, or even as a mask of Tutankhamun.

The sense of my present thus proves to be the fulfillment of my historicity. Everything I apperceive derives its sense from the accomplishments of my foregoing experience: to see the world already implies having learnt to see it.[9] Each and every world-experience thus proves to be historical through and through. The world thereby reveals itself not as a fixed formation, but rather as an accomplishment that is always on its way, always in the process of constitution.[10] The notion of the world as the *whereto* of experience is an expression of this open-ended historical dimension.

Thus this third notion of the world stands out from the other two as a notion that takes into account the world's concrete manifestation. But if so, then why was it necessary to establish a bond that would tie it to modalized experience? Not only does this bond run the risk of misportraying this notion by suggesting that it pertains only to the experience of the new or the uncommon. As we now see, each and every experience, be it modalized or unmodalized, is an experience that takes place within the world as the *whereto* of experience. Faced with such a realization, can the foregoing analysis of modalization still retain its credibility?

Let us turn back to the brief allusion I introduced in the beginning of this section. The allegation that the world as the *whereto* of experience originates in modalization should be conceived as a *genetic* claim. By this I mean that even though experiences, which can be phenomenologically accounted for in terms of a straightforward fulfillment of the anticipated sense, take place within the world as the *whereto* of experience, their origins point back to modalization. Only through modalization can one make sense of the origination of new types of experience: my anticipated horizon of sense must undergo a frustration if within it the unanticipated is to emerge. It is such an emergence of the unanticipated that crystallizes into a new type of experience.

Such an emphasis on modalization might appear suspect and one might be willing to retort that it does not take into account the amplitude and the fluidity of horizon-consciousness. There is no need for consciousness to undergo modalization in order to realize that, for instance, what one hears is a new genre of music, or what one sees is a new type of architecture. The amplitude of the horizon in such and similar cases is sufficiently inclusive to justify the emergence of new types by way of specification rather than modalization. Yet such a response is by far not conclusive because it leaves unclarified how consciousness has generated such general types of

---

[9] As Aguirre formulates it, "Das Erworbene ist das, was sich im Verlaufe aller meiner Erfahrungen als fester Besitz niedergeschlagen hat, es ist Widerspiegelung meines gesamten intentionalen Lebens als eines *Geschehens* der Erfahrung, es ist Widerspiegelung meiner eigenen *Geschichte*. Das Erworbene ist Geschichte, ist *Tradition*." (Aguirre 1970, 156)

[10] See in this regard *Mat. VIII*, 390.

experience. We therefore need to see that the consciousness of generalization and specification itself emerges on the ground of modalized experience. For such a purpose, we need to take a closer look at the origins of typifying consciousness.

12.3 The Origins of Typifying Consciousness

Thus the foregoing reference to the emergence of new types of experience brings us to a new problem. So as to render this problem more precise, let us turn to Husserl's remarks in B I 14 VI:

> If I get to know an unfamiliar animal, I do not just acquire knowledge of this individual, which I can always remember again and to which I can return at will perceptually, by renewing the acquired knowledge in a particular direction. Whenever I come across an individually different animal that I have never seen before but which belongs to the same type, I grasp it immediately as an animal of this type, that is, I apperceive it with all the characteristics, which I have earlier actually acquired, quite immediately through an "apperceptive transference of sense" on the basis of analogy. *Could I in general immediately experience any object in the world*, immediately and before any individual knowledge of this particular object, *if I did not have an antecedent framework of generally typical acquaintance, to which everything individual has to adapt itself?*[11]

That is, if consciousness immediately apperceives each and every appearance as an appearance of a particular type of objectivity, and if the type of objectivity in question has itself emerged on the basis of anterior appearances, then are we not compelled to concede that the phenomenological account of the genesis of types of experience leads to an infinite regress? In the manuscript under consideration, Husserl formulates this problem in terms of a peculiar kind of dialectic that emerges between what he terms *Habe* and *Vorhabe.* As he puts it, "Every perception is a self-having (*Selbsthabe*) and at the same time fore-having (*Vorhabe*)."[12] Now on the one hand, "every having (*Habe*) presupposes a fore-having (*Vorhabe*)"[13] yet on the other hand, "the employment of a fore-having presupposes a having."[14] Such a curious interplay of these two central elements of intuitive consciousness leads to the

[11] "Lerne ich ein fremdartiges Tier kennen, so erwerbe ich nicht nur eine Kenntnis dieses Individuums, als dieses, dessen ich mich immer wieder erinnern kann, zu dem ich wahrnehmend nach Belieben wieder zurückkehren kann, Kenntnis erneuernd und in neuen Richtungen fortbildend. Sooft ich aber später einmal individuell Anderes, ein nie gesehenes Tier derselben Art begegne, fasse ich es im ersten Blick als Tier dieser Art auf, d.h. ich apperzipiere es mit all den Eigenschaften, die ich im früheren mir wirklich erworben habe, ganz unmittelbar durch 'apperzeptive Übertragung' auf Grund der Analogie. *Könnte ich nun überhaupt irgendein Objekt der Welt unmittelbar erfahren*, unmittelbar und schon vor einer individuellen Kenntnis desselben erst schaffenden Tätigkeit gegeben haben, *wenn ich nicht schon früher den Rahmen der allgemeintypischen Bekanntheiten hätte, in die alles Individuelle sich einzufügen hat*?" (Hua XXXIX, 442–443)

[12] "Jede Wahrnehmung ist Selbsthabe und zugleich Vorhabe." (Hua XXXIX, 438)

[13] "Jede Habe setzt eine Vorhabe voraus." (Hua XXXIX, 438)

[14] "Der Anfang der Handlung, der Einsatz der Vorhabe, setzt voraus eine Habe." (Hua XXXIX, 440)

question, formulated in the title of this manuscript: "how is the original acquisition of the world possible?"[15] Yet this manuscript provides only a phenomenological description of the paradox itself; its resolution is here not to be found. And thus on the cover of this manuscript we find Husserl's inscription: "the problem is adequate; its presentation is unusable."[16]

One might be tempted to interpret this inscription as Husserl's direct admission that with this issue, we touch upon one of the limits of Husserl's phenomenology. However, there are good reasons to suggest that this inscription should not be understood as a reference to genetic investigations at large, but only as a reference to this particular manuscript. One could substantiate such a suggestion on the basis of those works, which Husserl wrote prior to the manuscript in question. I am referring to Husserl's analysis of *association* as presented in his *Vorlesungen zur passiven Synthesis*—a lecture course Husserl gave repeatedly in the 1920s.

Both in these lectures and in *Experience and Judgment*, Husserl identifies the syntheses of inner time-consciousness as a necessary presupposition of passive genesis in general, and of association in particular. Given Husserl's general determination of association as "something reminds of something else"[17] or "the present reminds of the past,"[18] one might be willing to interpret such a presupposition in terms of a pregivenness of a temporal field, within which associative fusions can occur. Yet such an interpretation, although not incorrect, does not exhaust the presupposition in question. More importantly, "time-consciousness is the original seat of the constitution of the unity of identity in general" (EJ, 73).[19] That is, time-consciousness is a fundamental presupposition of association not only because it provides a field within which associative concatenations can occur, but even more fundamentally, because it provides consciousness with a "something in general," with "proto-objectivity," "proto-phenomena," or in the words of *Experience and Judgment*, with "prominences" (*Abgehobenheiten*). Thus the presupposition of time-consciousness embraces "the unity of a prominent object and of a plurality of prominences as a simple multiplicity of coexistence and succession" (Hua XI, 134).[20]

From a static point of view, prominences are not given as a melee of chaotic data; they rather appear within a pregiven field of sense, which in its broadest determination is the surrounding world. Yet genetically, such a pregivenness of a field manifests itself as a more rudimentary accomplishment of consciousness. More particularly, it emerges from the *associative structures* of consciousness.

So as to account for the emergence of these structures, it is essential not to overlook the distinction Husserl draws between objectivities and prominences.

[15]"Wie ist ursprüngliche Erwerbung der Welt möglich?" (Hua XXXIX,438)

[16]"Das Problem genügt, Darstellung unbrauchbar." (Hua XXXIX, 862)

[17]"Etwas erinnert an etwas." (EU, 78)

[18]"Das Gegenwärtige erinnert an das Vergangene." (Hua XI, 118)

[19]"Das Zeitbewußtsein ist die Urstätte der Konstitution von Identitätseinheit überhaupt." (EU, 75–76)

[20]"[...] die Einheit eines abgehobenen Gegenstandes [und] Mehrheitenen von Abgehobenheiten als bloße Vielheiten der Koexistenz und Sukzession." (Hua XI, 134)

While objectivities are accomplishments of associative syntheses, prominences play the role of a fundamental presupposition of these accomplishments. Put otherwise, while objectivities are always already given within an anticipatory horizon of sense, the most rudimentary manifestation of prominences emerges in the absence of such anticipations, at least in regard to the *content* of what is anticipated. *Prominences are unities of affection that have not yet taken on the content of the horizonal framework.* Husserl's identification of association as the "universal principle of passive genesis" (Hua I, 82) must be therefore understood as an intimation that associative syntheses account for the manner in which the givenness of prominences gives rise to the anticipatory horizons of sense, or more particularly, to different *types* of objectivities.

Husserl identifies syntheses of *homogeneity* and *heterogeneity* as fundamental associative syntheses. The first one establishes an associative bond between different prominences in terms of affinity or similarity, while the second one—in terms of contrast or difference. Husserl privileges the syntheses of homogeneity over those of heterogeneity, as can be best witnessed in his analysis of mediate and immediate association (Hua XI, §26, EU, §16). Husserl asserts, emphasizing each and every word: "*but all immediate association is an association in accordance with similarity*" (EJ, 75).[21] *It is precisely the immediate syntheses of homogeneity that account for the genesis of the different types of objectivities.*

In its most elementary form, the type emerges in virtue of a synthetic connection that association establishes between two prominences. Husserl's privileged example in both *Lectures on Passive Synthesis* and *Experience and Judgment* pertains to the experience of colors. What from a static perspective manifests itself as an appearance of a red or green color calls for a genetic explanation, which establishes that my recognition of this appearance as red or green is rendered possible in virtue of an associative congruence (*Deckung*) between at least two prominences. Thus what emerges in experience as red or green can obtain such a sense due to the reawakening a foregoing experience, which shares with the present one its essential features. Due to such "associative reminders," which Husserl calls "etwas erinnert an etwas," consciousness is in possession of a particular type of "redness" or "greenness." One should not, however, overlook that such a genesis of types admits of generalization and specification. Thus my consciousness of what is red or green undergoes transformation with subsequent experience. Types of objectivities thereby reveal themselves as irreducibly malleable; or as Dieter Lohmar puts it, "types are downright voracious" (Lohmar 2003, 117).[22]

[21] "*Aber alle unmittelbare Assoziation ist Assoziation nach Ähnlichkeit.*" (EU, 78)

[22] Consider the example Dieter Lohmar employs in his account of the "beginning of experience": "the experiential history of almost all people begins with sensations like the characteristic taste of mother's milk and the warmth which is regularly connected with it. One is not likely to claim that an object in the sense of a conceived substrate with its properties is constituted through the connection between the two sensations. However, these sensations constitute a permanent synthetic unity by virtue of their reference to each other. We can refer back to this unity in further experiences through typifying apperception of the same lived experience. Either presentation can awaken the other as expectation and guide the constitution of the lived experience on the basis of the given sensations." (Lohmar 2003, 116)

We thereby see how association accounts for the genesis of typifying consciousness. My foregoing claim regarding the role that modalization plays in the phenomenological account of the origins of types of objectivities thereby reveals itself as a ramification of a more fundamental structure. It is thus necessary to render the foregoing analysis more precise: unmodalized experience continuously generates new types of experience; yet it can do so only because foregoing experience has given rise to new types through modalization. Modalization itself, however, is not the ultimate source of typifying consciousness, since it itself presupposes more rudimentary associative syntheses. Moreover, as I have argued in the earlier sections, associative syntheses themselves rest upon the syntheses of inner-time consciousness.

12.4 The *Whereto* of Experience and the Science of the Life-World

My foregoing analysis exposed the methodological framework that legitimates the conception of the world as the *whereto* of experience. It did not, however, account for the principal significance of this notion for the world-problematic as a whole in Husserl's phenomenology. Husserl has expounded the trenchant consequences that this notion entails in his analysis of the *life-world* in the *Crisis*. While a thoroughgoing analysis of this notion lies beyond the scope of this work, it is nonetheless necessary to show that the world as the *whereto* of experience bespeaks that dimension of world-experience without which no science of the life-world would be possible. By showing why such is the case, I will bring my analysis of the world-horizon as the *wherefrom*, the *wherein*, and the *whereto* of experience to completion.

At first glance, my reference to science might appear unwarranted. After all, I have consistently spoken of the world as it is latently (the *wherefrom*) or patently (the *wherein* and the *whereto*) given to *intuition*, while the world as scientifically analyzed is from the start claimed to be the world as given to *thinking*. Thus while the intuitive world is the correlate of *experience*, the scientific world is the correlate of specific *methods*. The phenomenological inquiry into world-experience derives its sense from this opposition. In contrast to the scientific attitude, which compels us "to take for *true being* what is only a *method*" (*Crisis*, 51),[23] phenomenology's pivotal task is that of disclosing the world's self-manifestation as it transpires prior to the adoption of the methodological standpoint that science inaugurates. Yet besides bestowing phenomenology with pertinence, such a coarse opposition also severely curtails the significance of phenomenological claims, for it indicates the unspoken acceptance of two truths, seemingly unrelated and indifferent to each other: "on the one side, everyday practical situational truths … on the other side, scientific truths" (*Crisis*, 132).[24]

---

[23]"Das Ideenkleid mach es, dass wir für *wahres Sein* nehmen, was eine *Methode* ist." (Hua VI, 52)

[24]"Auf der einen Seite die alltäglich-praktischen Situationswahrheiten[…]. Auf der anderen Seite die wissenschaftlichen Wahrheiten[…]." (Hua VI, 135)

First and foremost, such a crude opposition is hazardous in that it insinuates that the experiential dignity of phenomenological truths goes hand-in-hand with their scientific irrelevance. This opposition is no less problematic in that it ultimately cannot be justified on phenomenological grounds. On the one hand, the scientific truths are not utterly independent from subjective-relative truths; on the other hand, the truth of experience is not so self-contained that it could be claimed to be autonomous in regard to scientific practices. Husserl's analysis of science's dependence upon the prescientific world leads to the recognition of the life-world as a fractional problem that pertains to objective sciences (Hua VI, §33). Even more fundamental is the recognition that sciences themselves play a role within the life-world: such a state of affairs leads to the acknowledgment of the life-world as a *universal* problem within which sciences themselves constitute only a significant fragment (Hua VI, §34f). A gripping dialectic is at play between these two notions of the life-world. So as to address it, let us begin with the life-world as it manifests itself as a fractional problem within objective sciences.

Even though science establishes its claims in direct opposition to how the world reveals itself to prescientific experience, it cannot help but presuppose the pregivenness of the surrounding world from which it springs. The presupposition in question is not exhausted by the realization that science is a human accomplishment that takes its departure from the pregivenness of the life-world. In its everyday practices, science cannot help but continue to make use of those dimensions of sense, which belong to the world's prescientific manifestation. "For example, for the physicist it is the world in which he sees his measuring instruments, hears time-beats, estimates visible magnitudes, etc.—the world in which, furthermore, he knows himself to be included with all his activity and all his theoretical ideas" (*Crisis*, 121).[25] These inescapable references to the prescientific world indicate the constant role that the life-world plays within scientific practices.

To problematize the life-world within such a context is to thematize it as an *auxiliary* and *fractional* problem that falls within the general framework of objective sciences. The analysis of the life-world within such a context would face the task of clarifying the tacit presuppositions operative in scientific undertakings. If carried out successfully, such a phenomenological analysis would account for the sense and validity of science's continuous hearkening back upon the life-world; the phenomenological analysis of the life-world would thereby supply the sciences with a full grounding ("volle Begründung" [Hua VI, 125]) which it lacks, due to the coarse opposition between "two truths." "Thus the problem of the life-world … is only a partial subject within the above-designated whole of objective science" (*Crisis*, 122–123).[26]

[25]"Z.B. für den Physiker ist es die, in der er seine Meßinstrumente sieht, Taktschläge hört, gesehene Größen schätzt usw., in der er sich zudem selbst mit all seinem Tun und all seinen theoretischen Gedanken enthalten weiß." (Hua VI, 123–24)

[26]"So ist das Problem der Lebenswelt […] nur ein partielles Thema innerhalb des Ganzen der objektiven Wissenschaft." (Hua VI, 125)

Yet a functional account of the life-world as a fractional problem of objective sciences itself can only amount to a fractional account of the life-world. Such an account can only bring to light those dimensions of sense that perform a certain role in scientific practices. Clearly, one would be mistaken to identify those dimensions of sense with the life-world as a whole. But if so, then the life-world continues to hide even when it is analyzed as a fractional problem of objective sciences. Such a realization calls for a new approach to the life-world problematic.[27]

Before clarifying this new approach, Husserl provides us with a few helpful hints: he remarks that the task of the new approach consists in uncovering a "third dimension" and refers to the difficulties of obtaining the "right balance" (*richtige Fassung* [Hua VI, 126]).[28] Instead of reserving the English "comprehension" for the German "Fassung," I choose a less common translation "balance," because the problem we face now is precisely that of finding the proper mediation between the above-mentioned extremes: on the one hand, we need to avoid drawing a distinction between the world of sciences and the life-world in too coarse a way; on the other hand, we should be wary of not reducing the life-world to a fractional problem of objective sciences. The discovery of the right balance between these extremes must procure an approach that has the means of accessing the full problematic of the life-world. The latter, taken in its full and concrete manifestation, will prove to be the "third dimension" to which Husserl is alluding in this passage.

Within the context of such difficulties, that dimension of the world-horizon which I have characterized as the *whereto* of experience obtains eminent significance. Just as new types of experience do not only emerge from the world but also belong to it, so objective sciences themselves are subjective formations and as such they do not only presuppose the life-world as their ground but also belong to its full concreteness. "[The scientific] theories, the logical constructs, are of course not things in the life-world like stones, houses, or trees .... But this or any other ideality does not change in the least the fact that these are human formations, essentially related to human actualities and potentialities, and thus belong to this concrete unity of the life-world, whose concreteness thus extends farther than that of 'things'"

[27]One could further suggest that the distinction between the life-world as a fractional and a universal problem is an internal critique that Husserl directs at his own notion of transcendental logic. Husserl's analysis of prepredicative experience has from the start been restricted to a specific function, viz., that of accounting for the tacit presuppositions that underlie judgment itself. If Landgrebe is correct in identifying prepredicative experience with the life-world (EU §10), then we need to conclude that Husserl's analysis of the life-world in *Experience and Judgment* and *Formal and Transcendental Logic* is not universal, but only *fractional*. But if so, then one needs to assert that the domain of passivity and receptivity is more robust than Husserl's works on transcendental logic lead to believe. In this regard, the investigations carried through in the *Crisis* unmistakably enrich Husserl's foregoing analyses.

[28]"Eröffnet sich mit dem, was zunächst als ein spezielles wissenschaftstheoretisches Thema auftritt, nicht am Ende schon jene 'dritte Dimension', somit im voraus dazu berufen, das ganze Thema objektive Wissenschaft [...] zu verschlingen? [...]. Allen voran drängt sich hier auf und muß erwogen werden: die richtige Fassung des Wesens der Lebenswelt und die Methode einer ihr angemessenen 'wissenschaftlichen' Behandlung, wo doch 'objektive' Wissenschaftlichkeit außer Frage bleiben soll." (Hua VI, 126)

(*Crisis*, 130).[29] Husserl thus explicitly speaks of an *indentation* (*Einrückung*) that science leaves upon the life-world (Hua VI, 133).[30] Much like the sea indents the coastline, thereby relentlessly transforming it in such a way that it is no longer obvious how its original figure can be preserved or sustained, so the continuous streaming-in of the sciences into the life-world also modifies its sense in such a way that the notion of the prescientific life-world is no longer immediately accessible. The recognition that sciences are not just *about* the world, but also *of* the world, i.e., that they do not only emerge from the life-world but also belong to it, bespeaks the *primacy of the practical* in Husserl's phenomenology—and not only because all activities have their ultimate motive in the surrounding world, but also because each and every activity streams back into the life-world, thereby transforming it, yet by no means eradicating it.[31] One can thus say that sciences, conceived as particular practices of sense-formation, stand on the ground of the world as the *wherefrom* and the *wherein* of experience; yet they simultaneously stream back into the world, which is rendered possible by the world-horizon as the *whereto* of experience.

It is thus precisely the acknowledgment of the world as the *whereto* of experience, which embraces scientific practices and accomplishments, that justifies Husserl's claim that the life-world, far from being a fractional problem, in truth is a universal philosophical theme. Its universality relies upon the acknowledgment that each and every accomplishment of consciousness both stems from the life-world and belongs to it. With the recognition of the universality of the life-world problematic, phenomenology obtains the "right balance" between the above-mentioned extremes; the "third dimension" it uncovers is the life-world, conceived in its full and concrete self-givenness. Thus on the one hand, the life-world is no longer opposed to the sciences, but rather entwines them within its own compass; on the other hand, the life-world is no longer a fraction of the sciences, but on the contrary, a universal field of sense-formations within which science is one domain among others.[32] The world as the *whereto* of experience thereby reveals itself

[29] "[Die wissenschaftlichen] Theorien, die logischen Gebilde, sind zwar nicht lebensweltliche Dinge wie Steine, Häuser, Bäume[...]. Aber diese wie jede Idealität ändert nichts daran, daß sie menschliche Gebilde sind, auf menschliche Aktualitäten und Potenzialitäten wesensmäßig bezogen, und so doch zu dieser konkreten Einheit der Lebenswelt gehörig, deren Konkretion also weiter reicht als die der 'Dinge.'" (Hua VI, 132–33)

[30] In the English translation, *Einrückung* has been paraphrased as a "pull": "If we cease being immersed in our scientific thinking, we become aware that we scientists are, after all, human beings and as such are among the components of the life-world which always exists for us, ever pregiven; *and thus all of science is pulled, along with us, into the – merely 'subjective-relative' – life-world*" (*Crisis*, 130–31, *my emphasis*—S.G.). Yet arguably, as my subsequent remarks are meant to suggest, such a translation does not bring out in full force what Husserl here is trying to elicit.

[31] "[...] und sind nicht alle Ziele, ob sonstwie in einem außerwissenschaftlichen Sinn 'praktische' oder unter dem Titel 'theoretisch' praktische, eo ipso mit zur Einheit der Lebenswelt gehörig, wofern wir sie nur in ihrer ganzen und vollen Konkretion nehmen?" (Hua VI, 134)

[32] "Demgegenüber erscheint nun das Problem der 'objektiv wahren' Welt bzw. der objektiv-logischen Wissenschaft – wie sehr und mit wie gutem Grunde es sich immer wieder entgegendrängt – als Problem von sekundärem und speziellerem Interesse." (Hua VI, 136)

as that dimension of the world-horizon, which legitimates the phenomenological analysis of the life-world.

As I have indicated in Sect. 10.1, it is not my task here to undertake a thorough investigation into all the difficulties associated with Husserl's analysis of the life-world. Yet one should not overlook that even though the life-world deepens a number of phenomenological themes addressed in this study in a variety of ways and uncovers new directions for phenomenological research, it nonetheless is consistently disclosed on the basis of the three dimensions of the world-horizon distinguished in this work. I have disclosed the fundamental framework of the world-problematic in Husserl's phenomenology on the basis of his works other than the *Crisis*, and it is my contention that such a strategy is not without its merits. On the one hand, it indicates the continuity of Husserl's thought and thereby shows that the problematic Husserl addresses in the *Crisis* is not to be seen in opposition to his earlier works, but rather as an enrichment of the foregoing investigations. On the other hand, such a strategy also indicates that the *Crisis* is not Husserl's "only word" on the world-problematic, i.e., that it is not an exhaustive account of the world and that in fact certain dimensions of the world-horizon are elsewhere scrutinized in a more refined way.

My analysis has been guided by the distinction Husserl draws between the horizons of objectivities and the world-horizon.[33] I addressed this distinction in a threefold way. My analysis thereby disclosed a threefold sense in which each and every objectivity is from the outset marked as *worldly*. First, it is worldly due to its constitutive genesis, which necessitates the acknowledgment of the world as the *wherefrom* of experience. Secondly, each and every objectivity is worldly because its givenness in intuitive fulfillment simultaneously brings to self-givenness the world as the *wherein* of experience. Finally, objectivities are also worldly in a third sense: even when their emergence breaks apart the anticipatory horizons of sense, the establishment of their new modes of givenness does not take place outside the world, but rather transforms and enriches the phenomenality of the world, conceived as the *whereto* of experience. We thereby see that the horizons that objectivities carry with them are not reducible to those dimensions of apperceptive sense, which co-determine the manner of their manifestation. Alongside the inner- and outer-horizons that are awakened with the emergence of objectivity, we are led to acknowledge the co-awakening of an essentially different kind of horizon, i.e., of the world-horizon, which belongs to, yet simultaneously remains transcendent to the horizons of objectivity. Due to its threefold determination, the world-horizon can indeed be characterized as *unique* (Hua VI, 146) precisely because it is both inseparable from, yet irreducible to the givenness of concrete objectivities.

The world, as I suggested early on, is the originary figure of the horizon in Husserl's phenomenology. The present investigation leads to the acknowledgment

[33]For Husserl's own formulations of the distinction in question, see A VII 9, p. 15, as well as Husserl's response to Fink's *Sixth Cartesian Meditation*, published as Appendix XII. See also the short manuscript published as an appendix to *Ideas I*, Hua III, 399–400.

that its adequate understanding rests upon the three dimensions of sense that it entails. It is my contention that the threefold determination put forth is not only intrinsic to and indispensable from the world-problematic in Husserl's phenomenology, but also irreducible from the phenomenality of the world.

References

Aguirre, Antonio. 1970. *Genetische Phänomenologie und Reduktion*. Der Haag: Martinus Nijhoff.

Fink, Eugen. 1988. *VI. Cartesianische Meditation. Teil 1. Die Idee einer transzendentalen Methodenlehre*. The Hague: Martinus Nijhoff.

Husserl, Edmund. 1950. In *Ideen zu einer reinen Phänomenologie und phänomenologischen Philosophie. Erstes Buch: Allgemeine Einführung in die reine Phänomenologie*, ed. Biemel Walter. The Hague: Martinus Nijhoff.

Husserl, Edmund. 1966. In *Analysen zur passiven Synthesis: Aus Vorlesungs- und Forschungsmanuskripten, 1918–1926*, ed. Fleischer Margot. The Hague: Martinus Nijhoff.

Husserl, Edmund. 1970. *The crisis of European sciences and transcendental phenomenology. An introduction to phenomenological philosophy*. Trans. David Carr. Evanston: Northwestern University Press.

Husserl, Edmund. 1973. *Experience and judgment: Investigations in a genealogy of logic*. Trans. James Churchill and Karl Ameriks. Evanston: Northwestern University Press.

Husserl, Edmund. 1976. In *Die Krisis der europäischen Wissenschaften und die transzendentale Phänomenologie. Eine Einleitung in die phänomenologische Philosophie*, ed. Biemel Walter. The Hague: Martinus Nijhoff.

Husserl, Edmund. 1999. In *Erfahrung und Urteil. Untersuchungen zur Genealogie der Logik*, ed. Landgrebe Ludwig. Hamburg: Felix Meiner Verlag.

Husserl, Edmund. 2006. In *Späte Texte über Zeitkonstitution (1929–1934). Die C-Manuskripte*, ed. Lohmar Dieter. Dordrecht: Springer.

Lohmar, Dieter. 2003. Husserl's type and Kant's schemata: Systematic reasons for their correlation or identity. In *The new Husserl*, ed. Welton Donn. Bloomington: Indiana University Press.

Ricoeur, Paul. 1965. Universal civilization and national cultures. In *History and Truth*. Trans. C.A. Kelbey. Evanston: Northwestern University Press.

Chapter 13
Conclusion

13.1 The Question of the Unity of the Horizon

My foregoing investigation aimed to offer a systematic account of the notion of the horizon in Husserl's phenomenology. It was by far not an arbitrary decision on my part to provide such an account by way of raising the question of the origins of the horizon. Such a decision issued from the realization that the bond that ties the horizon-problematic to the question of origins bespeaks the most characteristic feature of Husserl's notion of the horizon—a feature that continues to distinguish Husserl's investigations into the horizon-problematic from the reemergence of this theme in post-Husserlian philosophy. It is my contention that the question of origins is indispensable for disclosing the development and the direction of Husserl's analyses of the horizon, as well as for revealing the distinctiveness of Husserlian investigations in the context of alternative inquiries into the horizon-problematic.

I raised the question of the origins of the horizon in a threefold way, following as my guide the threefold manner in which the horizon lends itself to a general characterization. First, insofar as the horizon is a philosophical theme and a philosophical notion, to raise the question of its origins meant to inquire into the emergence of this theme in philosophy as a whole, and in Husserl's phenomenology in particular. This was the task that guided my analysis in the first part of this investigation. Secondly, one can understand the horizon as the horizon of consciousness, or more broadly, as the horizon of subjectivity. On such a view, the horizon is a peculiar configuration of the stream of experiences, and thus to raise the question of the origins of the horizon meant to inquire into the crystallization of experience. Such was my central concern in the second part of this work. Finally, the horizon is also a particular figure of givenness, i.e., it bespeaks a unique and highly complicated pregivenness of the world. In this regard, to raise the question of the origins of the horizon is nothing other than to inquire into the different manners of the world's manifestation. This was my central concern in the third part of this study.

Such a multileveled indissociability of the horizon from the question of origins allowed me to distinguish a plurality of senses of the horizon in Husserl's

S. Geniusas, *The Origins of the Horizon in Husserl's Phenomenology*, Contributions to Phenomenology 67, DOI 10.1007/978-94-007-4644-2_13,

phenomenology. Yet such a plurivocity of senses, from which the horizon-problematic is inseparable, gives rise to a new concern: should one not say that the horizon is an irreducibly equivocal theme? The central question I want to address in this concluding chapter is therefore that of the *unity of the horizon*.

There are four main reasons that call for such an inquiry. (1) The first reason pertains to a shortcoming of my foregoing investigation. Although throughout the foregoing analysis I consistently spoke of the *notion* of the horizon, I did not establish how the manifold senses of the horizon discriminated in this work obtain their overarching unity. (2) The second reason relates to an analogous limitation present in the available literature on the horizon-problematic. In the few available commentaries, the question of the unity of the horizon is never raised while the task of disclosing the rich variety of senses that the horizon embraces is privileged. Because of these two reasons, one is left at a loss as to whether one can still speak of the *notion* of the horizon in the singular. (3) The third reason pertains to what I have already characterized as the distinguishing feature of Husserl's notion of the horizon: being inseparable from the question of origins, the Husserlian notion of the horizon embraces the *subjective*, the *historical* and the *ontological* (i.e., "worldly") dimensions of sense. The otherwise diverse inquiries into the horizon in post-Husserlian thought share the assumption that neither phenomenology, nor for that matter any other philosophical tradition, has the means to disclose the overarching unity that embraces these presumably non-unifiable dimensions. (4) Finally, the fourth reason pertains to what I see as a specific philosophical style and a specific philosophical task that Husserl's horizon-problematic gives rise to: both the style and the task in question pertain to the unity of the horizon, which its inseparability from the question of origins already intimates. By disclosing what it entails, I will bring this analysis to completion.

I will begin by elucidating the first two reasons. Afterwards, I will take a rather curtailed detour into the notion of the *historical apriori* as articulated in "The Origins of Geometry." This apparent digression will procure the means to speak of the horizon as a unity of sense. Having shown how this apparently contradictory notion bespeaks the unity of the horizon, I will be in the position to address the remaining two questions.

13.2 The Apparent Equivocity of the Horizon

Does the preceding analysis not lead to the realization that the horizons can be spoken of only in the plural? Such a conjecture does not only rely upon what could be seen as an accidental failure on my part to account for the unity in question. The hypothesis is more precarious in that it stems from a much more radical source—from the suspicion that the horizon is not accidentally, but rather *necessarily equivocal*. The plurality in question appears to be not just accidental because the expression "the question of origins" employed as a guide of this investigation already embraces an irreducible ambiguity, which is the very same one sooner or later is led to rediscover at the heart of the horizon-problematic. Let us take a closer look at the ambiguity in question.

"The question of origins" embraces two different dimensions of sense. On the one hand, it bespeaks a peculiar openness of each and every theme toward its own sense-genesis. To subject a phenomenon to the question of its origins is to *unfold the sense that accrues in its temporal unfolding*; to disclose its origins is to reflectively articulate the plurality of senses that constitute its phenomenality. Yet on the other hand, to raise the question of origins also means to *enunciate the pregiven ground of sense from which each and every theme emerges*. Such a task necessitates that one leaves the phenomenon in question behind so as to disclose the seemingly *extraneous* sources that underlie its emergence. Clearly, these two dimensions of sense are not identical. While the first one, in Husserl's phenomenology, calls for an inquiry into the static and genetic configurations of *lived-experience*, the second one leads to the phenomenological disclosure of the phenomenality of the *world*. Clearly, if the phenomenality of consciousness and of the world can be said to be irreducibly different themes, then the question of origins must also be characterized as irreducibly equivocal.

It is this very irreducible duality of sense that we rediscover at the heart of the horizon-problematic. This duality reemerges in the framework of the horizon-problematic under the headings of those two themes that respectively guided the foregoing analysis in the second and the third parts of this work. The themes in question are those of *the horizon of subjectivity* and *the world-horizon*. While these notions in their own frameworks were capable of uniting the different senses of horizonality, thc relation between them remains abstruse. What is it, one is willing to ask, that unites them, besides the mere presence of the world "*Horizont*?" Just like the question of origins, so the horizon also appears to be an irreducibly equivocal theme, whose unity is at best metaphorical. It seems that the horizon is not a notion at all.

Thus in the confines of the foregoing investigation, the question of the unity of the horizon turns out to be the question of a possible unification of *the horizon of subjectivity* and *the world-horizon*. Such is the case because in the preceding investigation I have already answered the question of the unity of the horizon, *yet only in part*. Namely, I have already shown that all the other senses of horizonality distinguished in this work fall under these two umbrella-terms, thereby allowing one to contend that the question of the unity of the horizon is precisely the question that pertains to the relation between these two notions. So as to render such a claim compelling, let us briefly turn to Roberto Walton's "On the Manifold Senses of Horizonedness"—a work which arguably is the most penetrating inquiry into the problem of horizonality in Husserl's phenomenology available in the English speaking world.

The central task of this work, as its title already announces, is that of disclosing the plurality of senses that pertain to the notion of the horizon. Walton's analysis, therefore, is of significance precisely because it problematizes the possible unity of the horizon more forcefully than other investigations. Walton thematizes the notion of the horizon in terms of no fewer than seven oppositions. The oppositions in question are those of potentiality/actuality, patency/latency, relevancy/irrelevancy, background/foreground, acquaintedness/non-acquaintedness, determinateness/indeterminateness, and finitude/infinitude. Faced with such a rich variety of senses that pertains to the notion of the horizon, the possibility of its unity appears highly improbable. Yet the question of the unity of the horizon does not lie outside the scope of Walton's analysis. Walton himself indicates two ways in which the question of unity can be posed. First, in an Aristotelian fashion, Walton suggests that the notion of the horizon, much like Aristotle's substance, has both a primary sense and derived senses. He grants the opposition between potentiality and actuality primary and fundamental status, thereby conceding that the other six oppositions should in fact be seen as subdivisions. Thus if one were to disclose the unity of the horizon on the basis of Walton's analysis, one would be led to speak of it on the basis of the potentiality/actuality distinction.

As I have already indicated, this is not the only way in which Walton's analysis leads to the question of the unity of the horizon. Walton also speaks of the encasement and stratification of horizons and suggests that such an *Ineinanderschachtelung* can generate the unity in question. Walton does not establish this hypothesis by means of providing a detailed basis the claim itself calls for; he rather presents it in the form of a conjecture:

> The temporal horizon can be first considered with regard to inner-time consciousness, secondly in the light of generative development … thirdly as "overmonadic or intermonadic time of higher level" (Hua XI, 582) or a "spiritual formation of the human being," and finally according to the imperishability or omnitemporality of ideal objects pertaining to the tradition of philosophy and science. The intersubjective horizon, which is first an open horizon of possible others, comes to be secondly a primal generativity with its home horizon, then a "higher-order generativity," and finally a rational and universal generativity. The world-horizon, which is first an indeterminate universal horizon, is secondly conceived as the "earthly horizon" (Hua XV, 206), i.e., the universal territory of all territories with the sky above, then as a life-world that develops a historical tradition on a territory, and finally as the world in itself that is thematized by philosophy and science through ideation and idealization. (Walton 2003, 16)

Clearly, the stratification in question is of decisive significance for Husserl's notion of the horizon, and as Walton himself remarks, it is this stratification that, unfortunately, is overlooked in post-Husserlian phenomenology. Motivated by similar concerns, I was led in the foregoing analysis to speak of the *plasticity* of the horizon in Husserl's phenomenology. Yet in the face of the voiced suspicion regarding the unity of the horizon, the attempt to obtain the oneness in question by means of disclosing the different strata can hardly be successful. For the suspicion continues to linger: is the stratification in question truly a stratification *of the horizon* and not of different themes or phenomena which only by chance are given one and the same name? Wherefrom comes the assurance that the encasement in question truly brings into relief different figures or dimensions *of the horizon*?

If Walton's analysis is to provide answers to these questions, they must stem from the first manner in which the horizon was said to be unified. We are thereby led back to the potentiality/actuality distinction. Yet this distinction, I would like to suggest, is already plagued with the duality of *the horizon of subjectivity* and *the world-horizon*. This opposition is more rudimentary than that of potentiality and actuality because the latter opposition, on phenomenological grounds, is itself already given as two-sided: the horizonal potentiality, as we have already seen in the first part of this work, is from the very start ambiguous in that it simultaneously stands for the potentiality of lived-experience and the potentiality of appearances.[1] Taken in its more robust fashion, this ambiguity is nothing else than the ambiguity of subjectivity and the world. Thus Walton's analysis of the horizon leads to the same equivocity of horizonality we have already seen arising in this work: how can the horizon be both the horizon of subjectivity and the world-horizon?

13.3 The Historical Apriori and the Unity of the Horizon

So as to address this question, let us turn to the notion of the *historical apriori* as thematized in Husserl's "Origins of Geometry." The most prominent interpretations of this work acknowledge that the notion of the horizon plays a decisive role in Husserl's analysis. Merleau-Ponty's commentary on this work culminates in the realization that "the theme of φ <"philosophy"> is the horizon of horizons" (Merleau-Ponty 2002, 67); and as Derrida formulates it in his *Introduction*, "the notion of horizon is decisive here: 'horizon-consciousness,' 'horizon-certainty,' 'horizon-knowledge,' such are the key concepts of the *Origin*" (Derrida 1989, 117). However, while these commentaries attest to the centrality of the horizon in Husserl's essay, I would nonetheless contend that neither of them discloses its full relevance. That such is the case can already be suspected on the grounds that Merleau-Ponty and Derrida have something entirely different in mind when they utter the above-mentioned phrases. For Merleau-Ponty, the centrality of the horizon ultimately stands for the pregivenness of the world, and thus the recognition of its significance allows for a transition from the analysis of this work to that of the "Foundational Investigations of the Phenomenological Origins of the Spatiality of Nature." For Derrida, the centrality of the horizon bespeaks the specificity of the notion of transcendental history in Husserl's phenomenology. These lines of interpretation can be reiterated in terms of the key-notions employed in the present investigation: while Merleau-Ponty sees the centrality of the horizon in terms of the *world-horizon*, Derrida interprets it as the *horizon of subjectivity*. Yet arguably, the centrality of the horizon in this text is precisely such that it embraces both lines of analysis. Let us turn to a brief consideration of the *Origin of Geometry* precisely because it holds the promise of revealing the unity of the two notions of horizonality.

The unity in question can be disclosed through Husserl's analysis of the *historical apriori*. The latter notion embraces a similar two-sidedness that lies at the heart of the horizon. Yet a crucial difference need not be overlooked between them: the

[1] For this very same reason, I would argue that the distinction between the horizon of subjectivity and the world-horizon also underlies the other six oppositions which, as Walton shows, pertain to horizon-intentionality.

duality in the case of the historical apriori neither does, nor even can appear as reducible to an equivocity. Rather, as we will soon see, this duality proves to be ineradicable from the *notion* of the historical apriori. We thus need to turn to the historical apriori so as to see how an irreducible two-sidedness that constitutes its sense does not disrupt its unity. Having established that such is the case, we will be in the position to raise the hypothesis that the historical apriori intimates a possibility that the notion of the horizon can also preserve its cohesion despite the plurivocity of senses from which it is inseparable.

Yet before describing the duality in question, I must interrupt this analysis with a seemingly weighty objection. At first glance, the notion of the historical apriori appears no less self-contradictory than that of wooden iron. One needs to ask: first, does the notion of the apriori not indicate what is pre-experiential and only subjective? And secondly, does the domain of history not presuppose an intersubjective and experiential basis? In virtue of such a seemingly radical breach between the historical and the apriori, Husserl's neologism appears to be nothing more than an attempt to bring into unity what by definition evades unification.

However, while this objection would hold its weight if the expression "historical apriori" was uttered by Descartes or Kant, it can be easily answered in the framework of Husserl's phenomenology. As Heidegger points out in his *History of the Concept of Time*, one of the greatest merits of Husserl's phenomenology is that of liberating the apriori from its restriction to what is merely subjective.[2] The apriori, as it appears in Husserl's works, is *indifferent* to the distinction between the subjective and the objective and it is precisely this indifference that justifies, at least in part, the employment of the phrase "historical apriori." For a full justification of this expression, one must also take into account that for Husserl, as Ricoeur has emphasized so forcefully,[3] history is necessarily a *history of the Idea*. A mere succession of facts is not yet history: while the former is largely indifferent to what makes its succession meaningful, history is possible only as a unity of sense, and as such, according to Husserl, it is possible only as a history of the Idea. For these two reasons, history lends itself to the disclosure of its apriori structures. Put briefly, the notion of the historical apriori embraces the essential structures and the necessary resources of the historical unfolding.

Yet by vindicating the historical apriori from the mentioned objection, one should be cautious not to overlook the tensions from which this notion is inseparable. The *temporal precedence* is one of the central senses of the apriori that is just as significant for Husserl as it is for Descartes or Kant. Thus as far as the historical apriori is *apriori*, it must "precede history"; yet as far as it is *historical*, it must already lie within history. How can the historical apriori be both historical and pre-historical? I would like to suggest that Husserl's answer to this problem lies in the above-mentioned irreducible two-sidedness of the historical apriori. Let us turn to this issue in more detail.

---

[2] See Heidegger (1979, §8).

[3] See Ricoeur, "Husserl and the Sense of History," in Ricoeur (1967), pp. 143–175.

First, the historical apriori is historical precisely to the degree that this notion depicts the structure of the historical unfolding. Husserl qualifies this structure in the following way: "history is from the start nothing other than the vital movement of the coexistence and the interweaving of original formations and sedimentations of sense" (*Crisis*, 371).[4] Yet secondly, the historical apriori is pre-historical inasmuch as it depicts the irreducible structure of the pregiven world, which remains unaltered throughout the historical unfolding. The insight into the invariant structures of the world's pregivenness is what accounts not only for the emergence of history, but also for the subsequent possibility of uncovering the sense of history from its sedimentations. Put in the language employed in the *Origin of Geometry*, the insight into the invariance of the world's pregivenness is the origin of geometry as well as the subsequent possibility to uncover its original motivation despite the multitudinous layers of historical sedimentation.

This two-sidedness of the historical apriori is ineradicable because both its historical and the pre-historical dimensions prove to be inseparably bound to each other. On the one hand, the interweaving of sense-sedimentation and sense-formation is possible only on the grounds of a more rudimentary horizon of sense-pregivenness: the sense must first be given so that it could be lost; it must first emerge so that it could undergo sedimentation. Moreover, for the absence of sense to have weight upon us, we must already have an inkling of what is being missed or what is being covered over. For Husserl, such a privation is directly experienced as a tension between the claims of science and the direct givenness of the world. Thus the pregivenness of the invariant structures of the world is an irreducible dimension of sense of the historical apriori: without presupposing its "pre-historical" dimension of sense, the "historical" dimension would lose its basis. Yet on the other hand, the original insight into the structures of the world's pregivenness is nothing outside its historical unfolding. The *pre*-historical is not *non*-historical: while the latter prefix can be considered an empty negation of history, the former is inseparably bound to history. Just like the origins of any tradition are meaningful only in relation to the tradition in question, so the pre-historical can have only a historical sense. Or as Derrida succinctly puts it in his *Introduction*, "the Idea is nothing outside history but the sense of all history …. The Idea is not an Absolute that *first* exists in the plenitude of its essence and then descends into history or becomes disclosed in a subjectivity whose acts would not be intrinsically indispensable to it" (Derrida 1989, 142). Thus the "pre-historical" sense of the historical apriori also presupposes the "historical" sense.

We need not follow further the two-sidedness of the historical apriori because the account provided so far, despite its curtailed character, has already illuminated the necessary resources to raise anew the question of the unity of the horizon. Could one not say that the two-sidedness of the horizon is analogous to the two-sidedness

[4] "Geschichte ist von vornherein nichts anderes als die lebendige Bewegung des Miteinander und Ineinander von ursprünglicher Sinnbildung und Sinnsedimentierung." (Hua VI, 380)

of the historical apriori in that the plurivocity of its senses does not disrupt, but on the contrary, reveals its ineradicable unity? Such a hypothesis can be confirmed in four steps.

1. First of all, if the analogy is to hold, then just as the different dimensions of sense that pertain to the historical apriori remain incomplete when taken in isolation from each other, so the different senses of horizonality should also remain fractional in the absence of the unity of the horizon. The analyses undertaken in the second and the third parts of this work confirm such a hypothesis. While the regressive inquiry into the horizons of subjectivity in the second part led to the recognition that the horizonality of experience itself already rests upon a more rudimentary world-horizon, the analysis of the different senses of the world-horizon in the third part of this work culminated in the realization that the world cannot be reduced to its pregivenness in that it embraces dimensions of sense that continuously undergo an enrichment of sense in the course of experience.
2. Yet the notion of the historical apriori opens the way to the disclosure of the unity of the horizon not only because of such a rather abstract structural analogy. Both the different dimensions of sense that constitute the historical apriori place pointers to the unfolding of the horizonal unity. Thus secondly, one must concede that *historicity* plays a central role in the disclosure of the unity of the horizon. The structural analogy is patent: just as the historical apriori is both pre-historical and historical, so the horizon has both subjective and ontological dimensions of sense. Yet curiously enough, in its thematic unfolding, it is the dimension of *historicity* that renders these two dimensions of horizonality inseparable from each other. My foregoing analysis makes this clear: on the one hand, it was precisely the inquiry into the *historicity* (i.e., genesis) of experience that led to the realization that the horizons of subjectivity rest upon the pregivenness of the world-horizon. On the other hand, it was the discovery of the *historicity* of the world-horizon itself (the world-horizon as the *whereto* of experience) that compelled us to concede that the world-horizon, taken in its full concreteness, can manifest itself only in the unfolding of lived-experience.
3. Thirdly, and even more significantly, the horizon manifests the same kind of indifference to the distinction between the subjective and the objective as does Husserl's notion of the apriori. An analogy here is indeed in place: no less than the apriori, the horizon cannot be reduced to either the subjective or the objective. Just as no appearance is possible outside the temporal horizons of subjectivity, so no horizons of subjectivity are possible without that which appears to subjectivity; and the appearance in question is such that it ultimately leads back to its emergence from the world-horizon. Thus the horizon, much like the apriori, does not stem from the distinction between the subjective and the objective but rather renders such a distinction possible.
4. Now precisely because the "pre-historical" and the "historical" dimensions of the historical apriori in different ways point to the unity of the horizon, the analogy

between the unity of the horizon and of the historical apriori is not as empty as the first point made might suggest. Thus fourthly, we need to take into account that the different dimensions of sense of the historical apriori proved not only incomplete when considered separately from each other; rather, the slightest determinacy of either is possible only inasmuch as it already stands in relation, even if it is only tacitly presupposed, to the other dimension. Even in this regard, the historical apriori is analogous to the horizon: just as there is no horizon without subjectivity, so there is also no horizon without that which appears to subjectivity, which is ultimately the world. That is, the horizon is an irreducibly *correlative* notion.

Such a fourfold significance of the historical apriori for the establishment of the unity of the horizon thereby leads to a better grasp of what it means to qualify the horizon as a *figure of intentionality*. The searched-for unity of the horizon is thus itself intentional: subjectivity, taken abstractly in complete isolation from the world, has no horizons of its own; the world, conceived in its isolation from subjectivity, is just as horizoneless. The horizon can therefore be neither merely subjective, nor merely worldly, and only by means of entailing both dimensions does it reveal itself in its unity. Thus on the one hand, an inquiry into the horizons of subjectivity already carries with it the world-horizon in the background; and on the other hand, the disclosure of the world-horizon can only emerge in the framework within which a place is already reserved for the horizons of subjectivity. The irreducible two-sidedness of the horizon is therefore nothing other than the correlative manifestation of the horizon-intentionality. One thus needs to assert that the horizon obtains its unity not in spite of, but rather precisely due to its two-sidedness.

So as to qualify the horizonal correlation between the world and subjectivity in a more precise way, one needs to assert its *genetic* character. There are two reasons that underlie such a qualification. First, the elements of the correlation in question are themselves figures of the horizon. Secondly, the horizon is a specifically genetic theme, as we have already seen in the second part of this investigation. Thus any configuration of sense that qualifies the present horizon of subjectivity and the present horizon of the manifestation of the world already rest upon more rudimentary accomplishments of subjectivity and correlatively, upon a more rudimentary phenomenality of the world. *An inquiry into the horizon is therefore always already horizonal.* That is, the analysis of the horizon can only be such that as it renders one of its dimensions thematic, the horizon's other dimension of sense recedes into the background. Such a state of affairs necessitates one to reassert the significance of the question of origins for the horizon-problematic, for it is precisely by way of raising the question of origins that the inquiry into the horizon is led to the recognition of what, throughout the thematization, persists as a non-thematic dimension. One can thus say that the horizon, conceived in terms of its genetic correlation, reveals its unity in the wake of the origins-problematic.

13.4 The Origins of the Horizon and the Question of Subjectivity

What are the philosophical implications that stem from the foregoing analysis? With this question in mind, let me briefly turn to the final pages of Mikhail Bulgakov's *Master and Margarita*. Here we find a set of questions that are of paramount significance for the philosophical inquiry into the horizon-problematic. I would like to single out a few questions:

> What would the world look like without shadows? Shadows are thrown by people and things. There's the shadow of my sword, for instance. But shadows are also cast by trees and living beings. Do you want to strip the whole globe by removing every tree and every creature to satisfy your fantasy of a bare light? [5] (Bulgakov 2002, 404)

Echoing these questions, Husserl's late assistant, Eugen Fink, once remarked: "schattenlos erkennt allein der Gott," "only God knows the world in a shadeless, shadowless, way."

One could say that it is precisely the re-inscription of shadows back into the human universe that constitutes the central accomplishment of Husserl's analyses of the horizon. Yet curiously enough, for Husserl, the shadows that are irreducible from our experience of the world, derive from the concealed and sedimented accomplishments of subjectivity. For Husserl, the intuitive emptiness, which engulfs and determines any appearing phenomenon, is itself an accomplishment of subjectivity. For this reason, I would like to suggest that the most significant accomplishment of Husserl's analyses of the horizon lies in the realization that such an analysis can bring to light—and must bring to light—those dimensions of experience that are of paramount significance for our understanding of *who we ourselves are*.

Working on the basis of Husserl's analyses of the horizon, one can unfold a conception of subjectivity which in a number of important ways would complement those conceptions of subjectivity that have found a prominent place in contemporary philosophy. Indeed, the above-mentioned indissociability of the horizon from the question of origins contrasts sharply with the manner in which the problematic of the horizon reemerges in more recent philosophical perspectives. By inheriting the problematic of the horizon from Husserl's phenomenology, the more recent post-Husserlian traditions did not inherit the bond that ties it to the question of origins. Due to the absence of the problematic of origins, horizon-intentionality no longer manifests the unity of sense it obtained in Husserl's phenomenology. Of course, the horizon still remains a horizon of subjectivity and of the world. Yet the question of how subjectivity comes to obtain horizons that embrace concrete phenomena no longer elicits philosophical interest. If Gadamer's phenomenological hermeneutics can be characterized as the most elaborate inquiry into the horizon in

[5] "[…] как бы выглядела земля, если бы из нее изчезли тени? Ведь тени получяются от предметов и людей. Вот тень от моей шпаги. Но бывают тени от деревьев и от живых существ. Не хочешь ли ты ободрать весь земной шар, снеся с него прочь все деревья и все живое из-за твоей фантазии наслаждаться голым светом?" (Булгаков 2002, 404)

post-Husserlian thought, then this philosophical framework makes it patently clear that the horizon-problematic has become not only independent from, but also indifferent to the transcendental problematic of the origins of sense-formation. The problematic of disclosing the sedimented dimensions of sense has receded into the background so as to free the space for the progressive fusion of the horizons. And even though the question of self-understanding unquestionably remains a central concern for the hermeneutical inquiry into horizonality, it is no longer to be obtained by thematizing the genesis of that horizonality, which governs over the unfolding of subjectivity's transcendental life. The contribution that hermeneutics makes to the question of subjectivity through the horizon-problematic consists in a patient description of how subjectivity manifests itself *already within* a horizonal framework.

Of course, such a liberation of the horizon from the problematic of origins does not mean that the question of origins has itself become obsolete in phenomenology. This question retains its centrality in contemporary French phenomenology under the heading of the problematic of givenness.[6] Yet as we have already seen, when the question of givenness comes to the fore, it emerges in the absence of the horizon-problematic. To be more precise, the novel manner of its emergence is secured by its liberation from the horizonal framework of experience. One already encounters such a distantiation of the question of origins from the horizon-problematic in Heidegger's works, for arguably, it is precisely the need to raise the question of origins in a truly radical fashion that compels Heidegger to abandon both the notion and the problematic of the horizon to which he still clings in *Being and Time*. Jean-Luc Marion follows this path initiated by Heidegger, and thus the "third reduction" of which he speaks in *Reduction and Givenness* is meant to bring this separation of origins from the horizons to the point of no return: it is meant to secure a dimension of givenness in the domains that lie beyond transcendental subjectivity's reach and thereby, in the absence of horizonal configuration, to disclose the given "without limit, condition, or reserve" (Marion 1998, 16).

Let us set aside the questions concerning the breadth of horizonality when it is thematized in the absence of the question of origins; let us also suspend the questions regarding the possibility of inquiring into origins in the absence of the horizon. So as to address these questions in a compelling way, we would need to trespass the boundaries of Husserl's phenomenology. Such a task, although indeed of great philosophical interest, lies beyond the scope of this work. Yet it needs to be emphasized that the development of the horizon-problematic in post-Husserlian thought has neither rendered its bond to the question of origins illegitimate, nor has it established that a phenomenologically justifiable question of origins can emerge only on the other side of the horizon-problematic. The Husserlian perspective, therefore, remains to this day a viable philosophical alternative. In my concluding remarks, I would like to address the potency of this philosophical perspective by questioning the heritage with which Husserl's analyses of the horizon have left us.

[6] See in particular Jean-Luc Marion's *Reduction and Givenness* and *Being Given*.

Seen from the perspective of Husserl's phenomenology, the manner in which the horizonal investigations have been undertaken in hermeneutics remain deficient in that they overlook the question of the *emergence* of horizonality. Whether the horizonal framework of experience is thematized under the heading of *facticity* (Heidegger) or that of *history of effect* (Gadamer), these analyses take their departure in the *acceptance* of the horizonal framework that engulfs subjectivity without *genetically* problematizing its ineradicable presence. Thus Heidegger's rich description of Dasein's being-in-the-world provides an unprecedented account of our own facticity, yet without bringing into question the *emergence* of the world-horizon itself. Gadamer's no less brilliant account of the pregivenness of history and tradition also leaves the question of the origins of the horizon out of account. In fact, both Heidegger's and Gadamer's philosophical standpoints block access to the emergence of such a question, suggesting that the motivation that underlies the problem itself lies in the fictive notion of a *worldless* subjectivity. Yet is the question of the origins of the horizon really out of place within a phenomenological framework? Is subjectivity, which recognizes itself as being-in-the-world, really indifferent to the question of the genesis of sense-formation? And if the question of who we ourselves are is still to be considered philosophy's overriding concern, then should the question of how subjectivity obtains a horizonal framework be so single-handedly dismissed as meaningless and illegitimate? The manner in which Husserl's notion of the horizon shows itself as intentionally unified reveals that there is no necessity to choose between the pregivenness of the world-horizon and the question of the origins of the horizons. One is not only in perfect right to inquire into the genesis of the horizons while conceding the pregivenness of the *Welthorizont*; in fact, only within such an intentional framework, as the last section has shown, can the question of origins be posed. Thus the analysis undertaken in this work necessitates that the questions above be answered with an unequivocal *no*.

I would suggest that the most significant advantage of a Husserlian approach to the horizon-problematic consists in disclosing those dimensions of sense that, while they are overlooked in post-Husserlian philosophy, are of paramount significance for our understanding of who we ourselves are. By legitimizing the question of the origins of the horizon and considering this question the very problematic of the horizon, Husserl's phenomenology discloses concealed and sedimented accomplishments of subjectivity. One can highlight these hidden manifestations of our own selves with the help of an expression which, even though not employed by Husserl himself, is exceptionally well suited to characterize the specificity of his phenomenology. Through the horizon-problematic, we enrich our understanding of who we ourselves are by disclosing the *plasticity of experience*.

This expression is meant to suggest that horizons have their own genesis within our life. It is the malleable nature of experience itself that generates the horizons of sense that engulf appearing phenomena. Thus my foregoing analysis showed how each experience we undergo leaves a permanent imprint upon us. We saw how, while each experience "streams in," there is a sense in which it does not "stream away": we retain it as a *sedimented* sense; we retain it as our own *habituality*. We thus saw that it is experience itself that generates new types of experience, i.e., new horizons of sense, which in their own turn guide our subsequent experience. It is precisely due to sedimentations, and more generally, to habitualities, that we can have horizons as general frameworks of sense that embrace concrete appearances. Yet so as to uncover the logic that governs over the formation of our own habituality, it proved necessary to raise the question of the origins of the horizon. And as we saw, Husserl's patient pursuit of this question culminates in the recognition of the pregivenness of the world as the fundamental figure of the horizon. We thereby see why the horizon-problematic, interpreted through the question of its origins, is absolutely essential for our understanding of who we ourselves are: the horizon-problematic reveals that subjectivity is a "store" of sedimented accomplishments due to which the world is given to us not as a set of meaningless data but as a meaningful whole.

If the Delphic dictum "Know Thyself" can still be considered philosophy's abiding vocation, then the Husserlian problematic of the origins of the horizon discloses unseen features of our own selves as well as opens new avenues for the further pursuit of this ever-lasting philosophical enigma.

References

Булгаков, Михаил. 2002. *Мастер и Маргарита*. Москва: Издателсво АСТ.

Derrida, Jacques. 1989. *Edmund Husserl's origin of geometry. An introduction*. Trans. J. Leavey. Lincoln/London: University of Nebraska Press.

Heidegger, Martin. 1979. *Prolegomena zur Geschichte des Zeitbegriffs*. Frankfurt am Main: Vittorio Klostermann.

Husserl, Edmund. 1970. *The crisis of European sciences and transcendental phenomenology. An introduction to phenomenological philosophy*. Trans. David Carr. Evanston: Northwestern University Press.

Husserl, Edmund. 1976. In *Die Krisis der europäischen Wissenschaften und die transzendentale Phänomenologie. Eine Einleitung in die phänomenologische Philosophie*, ed. Biemel Walter. The Hague: Martinus Nijhoff.

Marion, Jean-Luc. 1998. *Reduction and givenness. Investigations of Husserl, Heidegger, and phenomenology*. Trans. Thomas Carlson. Evanston: Northwestern University Press.

Merleau-Ponty, Maurice. 2002. *Husserl at the limits of phenomenology*. Trans. Leonard Lawlor and Bettina Bergo. Evanston: Northwestern University Press.

Ricoeur, Paul. 1967. *Husserl: An analysis of his phenomenology*. Evanston: Northwestern University Press.

Walton, Roberto. 2003. On the manifold senses of the horizonedness. The theories of E. Husserl and A. Gurwitsch. *ch* 19(1): 1–24.

Index

S. Geniusas, *The Origins of the Horizon in Husserl's Phenomenology*,
Contributions to Phenomenology 67, DOI 10.1007/978-94-007-4644-2,

The manufacturer's authorised representative in the EU is Springer Nature Customer Service Centre GmbH, Europaplatz 3, 69115 Heidelberg, Germany. If you have any concerns regarding our products, please contact ProductSafety@springernature.com

Printed and bound by CPI Group (UK) Ltd, Croydon, CR0 4YY

30/09/2026

02225447-0003